Selected Titles in This Series

(Continued in the back of this publication)

Translations of
MATHEMATICAL
MONOGRAPHS

Volume 208

IWANAMI SERIES IN MODERN MATHEMATICS

An Introduction to Morse Theory

Yukio Matsumoto

Translated by
Kiki Hudson
Masahico Saito

American Mathematical Society
Providence, Rhode Island

Editorial Board

Morse理論の基礎

MORSE RIRON NO KISO
(AN INTRODUCTION TO MORSE THEORY)
by Yukio Matsumoto

Copyright © 1997 by Yukio Matsumoto
Originally published in Japanese
by Iwanami Shoten, Publishers, Tokyo, 1997
Translated from the Japanese by Kiki Hudson and Masahico Saito

2000 *Mathematics Subject Classification.* Primary 57–01;
Secondary 57R19, 57R65, 57R70, 57M25, 57M99.

ABSTRACT. This book aims at introducing Morse theory to undergraduate or basic graduate students. The emphasis is on Morse theory on finite-dimensional manifolds. The topics covered include Morse functions, handlebodies, handle decomposition of various manifolds and Lie groups, sliding and canceling handles, Poincaré duality, intersection forms, low-dimensional manifolds, and Kirby calculus.

Library of Congress Cataloging-in-Publication Data

Matsumoto, Y. (Yukio), 1944–
[Morse riron no kiso. English]
An introduction to Morse theory / Yukio Matsumoto ; translated by Kiki Hudson, Masahico Saito.
    p. cm. — (Translations of mathematical monographs, ISSN 0065-9282 ; v. 208)
(Iwanami series in modern mathematics)
Includes bibliographical references and index.
ISBN 0-8218-1022-7 (pbk. : acid-free paper)
1. Morse theory. I. Title. II. Series. III. Series: Iwanami series in modern mathematics.

QA331.M442713   2002
514—dc21   2001045751

Contents

Preface

In a very broad sense, "spaces" are objects of study in geometry, and "functions" are objects of study in analysis. There are, however, deep relations between functions defined on a space and the shape of the space.

For example, let us consider a line and a circle. Both are one-dimensional spaces. Identifying a line with the x-axis, there are functions which take as arbitrarily large values, such as

$$y = x^2, \quad y = x^3.$$

On the other hand, there exists no such function on a circle. For, any continuous function on a circle must take a maximum value somewhere on the circle (maximum value theorem). This way, we can distinguish a line and a circle by whether or not there are functions on them that take arbitrarily large values.

Morse theory is the study of the relations between functions on a space and the shape of the space. In particular, its feature is to look at the critical points of a function, and to derive information on the shape of the space from the information about the critical points.

Morse theory deals with both finite-dimensional and infinite-dimensional spaces. In particular, it is believed that Morse theory on infinite-dimensional spaces will become more and more important in the future, as mathematics advances further.

In this series, there are two books on Morse theory. This volume, "An Introduction to Morse Theory," describes Morse theory for finite dimensions. The other volume, "Geometric Variation Problems" by S. Nishikawa, deals with infinite dimensional aspects of Morse theory.

Finite-dimensional Morse theory has the advantage that it is easier to present the fundamental ideas than in infinite-dimensional Morse theory, which is theoretically more involved. Therefore, finite-dimensional Morse theory would be more suitable for beginners to study.

On the other hand, finite-dimensional Morse theory has its own significance, not just as a bridge to infinite dimensions. It is an indispensable tool in the topological study of manifolds. That is, we can decompose manifolds into fundamental blocks such as cells and handles by Morse theory, and thereby compute a variety of topological invariants and discuss the shapes of manifolds.

These aspects of Morse theory will continue to be a treasure in geometry for years to come.

This book was written with $\mathcal{A}_{\mathcal{M}}\mathcal{S}$-LaTeX, with which the author was unfamiliar. The author would like to express his gratitude to Professor Toshio Oshima at the Graduate School of Mathematical Sciences, University of Tokyo, for installing the program for him, as well as other help the author received.

Finally, the author would like to thank the editors of Iwanami Shoten, who were of great help in the publication of the book.

Yukio Matsumoto
February, 1997

Preface to the English Translation

This book was published in 1997 in Japanese by Iwanami Shoten, as a volume in the series "Foundations of Modern Mathematics." It is a great pleasure to me as the author that the English translation is published by the American Mathematical Society. I wish to express my gratitude by reflecting on the circumstances leading to the English translation.

It was November 1997 when I received a request for publication of the English translation from Professor Katsumi Nomizu at Brown University. Since then, up until a few months ago, I have been involved in administrative work for the Mathematical Society of Japan, and it seemed difficult to carry out the translation by myself, so I had to ask someone else. Dr. Kiki Hudson, who had been a colleague for over thirty years in the group of topologists in Tokyo, came to my mind at once. She was a topologist who was fluent in foreign languages. She already had experience in translating several Japanese books in mathematics into English, and vice versa.

Kiki accepted my request with a good grace. I recall that her work started in 1998. I was looking forward to seeing her translation completed, but I refrained from making contact with her, as I was afraid it might disturb her in her busy schedule. I regret this very much now. For I was surprised and deeply saddened by the unbelievable news that she passed away from cancer in September 1999. Several months before her death, Kiki called and told me that she had gotten hospitalized from a bad cold, but was feeling better at the time. I had never imagined that her illness was so grave.

After she passed away, her husband, Dr. Hugh Hudson, told me that she had finished her translation up to the beginning of Chapter Three, in spite of her serious illness. I was filled with sadness and gratefulness.

Hugh contacted Ms. Chris Thivierge at the AMS, and mailed Kiki's manuscript to her. I did not want to waste Kiki's great effort, and asked Dr. Masahico Saito at the University of South Florida to finish her work. Masahico was my student at the University of Tokyo. He, too, gladly accepted my request, and translated the rest of the book, paying attention to consistency throughout the book.

When the first draft was completed, an old friend of mine, Professor José María Montesinos-Amilibia at the Universidad Complutense de Madrid, kindly read through the manuscript, and gave me numerous comments. Furthermore, he pointed out a few errors in the original. I believe that José María's comments greatly improved the contents and exposition of the book.

The English translation of the book would have never been accomplished without the great contributions and help of the people I mentioned above. I would like to express my deepest gratitude to them.

With the translation in hand now, I appreciate, deeply from my heart, the old saying: "Men do not live alone."

Yukio Matsumoto
Tokyo, Japan
June, 2001

Objectives

The primary concern of Morse theory is the relation between spaces and functions. The center of interest lies in how the critical points of a function defined on a space affect the topological shape of the space, and conversely, how the shape of a space controls the distribution of the critical points of a function.

Morse theory of finite-dimensional manifolds is a powerful tool for the topology of manifolds, and offers a unified method to "visualize" manifolds with theoretical eyes. On the other hand, Morse theory for infinite-dimensional spaces clarifies the deep relations between variational problems and geometry, and is one of the basic principles of modern mathematics.

In this book, which deals with finite dimensions, we first introduce fundamental concepts such as critical points, the Hessian, and handle decompositions, with surfaces as examples.

These are generalized to higher dimensions in Chapter 2. The existence of enough Morse functions will be proved in this chapter as well.

In Chaper 3, handle decompositions associated with Morse functions are discussed in general dimensions, and the theory of handlebodies is developed. When we said that Morse theory offers a unified method to visualize manifolds, we had handlebodies in mind. Furthermore, we explicitly construct Morse functions on classical spaces such as projective spaces and Lie groups, and compute indices and the numbers of critical points. Moreover, the fundamental tools for dealing with handlebodies, such as sliding handles and canceling pairs, will be explained. This chapter is most essential.

In Chapter 4, we discuss the relation between handle decompositions and cell decompositions, and we see that homology theory is made easier to visualize by using handlebody structures of manifolds. For example, Poincaré duality is nothing but an algebraic expression of "turning a handlebody upside down." Also, it seems that discussions on intersection forms on manifolds can be made well-balanced between intuition and theoretical rigor by using handlebody structures.

Chapter 5 is devoted to low-dimensional (dimension 4 and less) manifolds. In low dimensions, handlebodies can be visualized explicitly by Heegaard diagrams and Kirby diagrams, which can be drawn on a piece of paper as framed links. Such concreteness makes us feel familiar with low-dimensional manifold theory, and at the same time, the relation to knot theory becomes evident immediately. It can be said that knot theory and low-dimensional manifold theory are almost the same subject. Both are very concrete and make us feel familiar, but they are not easy subjects. We realize the difficulty right away if we try to untie a tangled thread. (Imagine a closed circle C in 3-space, which is knotted in a very very complicated manner. Together with a specified integer n, a closed 3-manifold $M_{(C,n)}$ is assigned to the pair (C, n) via a Kirby diagram. Such a correspondence will convince the reader of the incredible complexity of the subject.)

Low-dimensional manifold theory is an area which is very actively studied today. The original plan was to make Chapter 3 and this Chapter 5 the main topics of the book, but unfortunately, due to page limitations, only fundamental notions are covered in Chapter 5.

In my opinion, the most interesting aspect of handlebody theory, either in low or high dimensions, is its geometric entity, which we feel as if we can touch and see with our eyes. I feel that, in Morse theory, we are not much different from children playing with wood blocks.

The purpose of this book would be more than half achieved if such a feeling is passed along to the reader.

CHAPTER 1

Morse Theory on Surfaces

Morse theory investigates how functions defined on a manifold are related to geometric aspects of the manifold. In Chapter 1 we investigate these relations for surfaces. Surfaces are easy to visualize, and all the essential points of the theory readily appear in the case of surfaces.

1.1. Critical points of functions

Let us consider a function $y = f(x)$ in one variable. We assume that both x and y are real numbers. Recall that a basic method for investigating the increase and decrease of a function f is to differentiate f, find the points x_0 for which $f'(x_0) = 0$, and then study the changes of the derivative $f'(x)$ around x_0. A point x_0 which satisfies

$$f'(x_0) = 0$$

is called a *critical point* of the function f. The points at which f takes its maximum or minimum values, and the inflection point of $y = x^3$, are examples of critical points.

The critical points of f fall into two classes according to the values $f''(x_0)$ of the second derivative of f. We call x_0 a *non-degenerate critical point* if $f''(x_0) \neq 0$, and a *degenerate critical point* if $f''(x_0) = 0$.

EXAMPLE 1.1. The quadratic function $y = x^2$ gives $y'' = 2$; so the critical point $x = 0$ of $y = x^2$ is non-degenerate.

For $n \geq 3$, the critical point $x = 0$ of the degree n function $y = x^n$ is degenerate. In fact, the second derivative $y'' = n(n-1)x^{n-2}$ of $y = x^n$ is zero at $x = 0$.

Looking at how the graphs of some functions meet the x-axis (see Figure 1.1), one sees that the graph gets somewhat closer with a higher degree of tangency to the x-axis at a degenerate critical point than it does at a non-degenerate critical point.

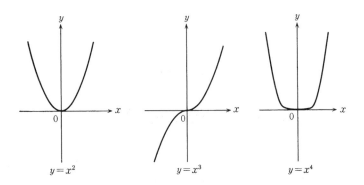

FIGURE 1.1. The graphs of $y = x^2$, $y = x^3$, and $y = x^4$

Another difference, between non-degenerate critical points and degenerate ones, becomes evident when one changes (perturbs) the function slightly. Consider the quadratic function $y = x^2$ and the cubic function $y = x^3$. Each has a critical point at $x = 0$. For $y = x^2$ it is a non-degenerate critical point and for $y = x^3$ it is a degenerate critical point. We perturb these functions by adding a linear function $y = ax + b$.

We find critical points of the perturbed function of $y = x^2$:

(1.1) $$y = x^2 + ax + b.$$

Differentiating y, we obtain

$$y' = 2x + a.$$

The point $x = -a/2$ at which the derivative vanishes is the only critical point of the function (1.1). Since its second derivative is $y'' = 2$, we see that $x = -a/2$ is a non-degenerate critical point of (1.1); that is, the critical point $x = -a/2$ appearing near the non-degenerate critical point $x = 0$ after perturbation is again non-degenerate.

What about degenerate critical points? The derivative of the function

(1.2) $$y = x^3 + ax + b$$

obtained by perturbing the cubic function $y = x^3$ is

$$y' = 3x^2 + a.$$

We set this equal to zero to get

(1.3)
$$x = \pm\sqrt{\frac{-a}{3}}.$$

The solutions (1.3) are not real solutions when $a > 0$, so the perturbed function has no critical point if $a > 0$. We lost a critical point after the perturbation.

When $a < 0$, the two values of x in (1.3) are real and hence we have two critical points. Since the second derivative $y = 6x$ of the function $y = x^3 + ax + b$ takes non-zero values at $x = \pm\sqrt{-a/3}$, these critical points are non-degenerate.

The degenerate critical point $x = 0$ of the original function $y = x^3$ vanishes $(a > 0)$ or splits into two non-degenerate critical points $(a < 0)$ depending on the way we perturb the function.

We conclude that non-degenerate critical points are "stable" and degenerate critical points are "unstable."

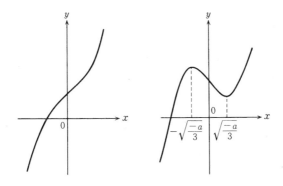

FIGURE 1.2. The graphs of $y = x^3 + ax + b$: $a > 0$ (left), $a < 0$ (right)

1.2. Hessian

We now move to a real-valued function

(1.4)
$$z = f(x, y)$$

of two variables, where x and y are both real numbers. We may think of a pair (x, y) of real numbers as a point in the xy-plane. In this way the function (1.4) becomes a function defined on the plane, which assigns a real number to each point in the plane. We can visualize the graph of this function in the 3-dimensional space with three orthogonal axes x, y, z.

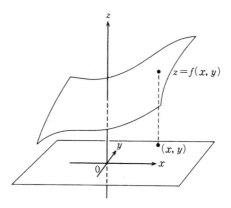

FIGURE 1.3. An example of a graph of a function of two variables

Such an example of the graph of a function of two variables is depicted in Figure 1.3.

DEFINITION 1.2. (Critical points of functions of two variables). We say that a point $p_0 = (x_0, y_0)$ in the xy-plane is a *critical point* of a function $z = f(x, y)$ if the following conditions hold:

$$(1.5) \qquad \frac{\partial f}{\partial x}(p_0) = 0, \quad \frac{\partial f}{\partial y}(p_0) = 0.$$

We assume in this definition that the function $f(x, y)$ is of class C^∞ (differentiable to any desired degree). Such a function is also called a C^∞-function, or a smooth function. In this book we only consider functions of class C^∞.

EXAMPLE 1.3. The origin $\mathbf{0} = (0, 0)$ is a critical point of each of the following three functions:

$$(1.6) \qquad z = x^2 + y^2, \quad z = x^2 - y^2, \quad z = -x^2 - y^2$$

(see Figure 1.4).

How do we best define non-degenerate and degenerate critical points for functions of two variables?

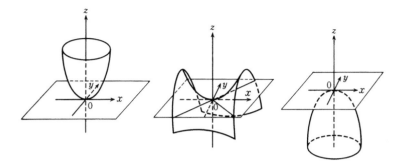

FIGURE 1.4. The graphs of $z = x^2 + y^2$, $z = x^2 - y^2$, and $z = -x^2 - y^2$, respectively from the left

After a moment of reflection the reader may be tempted to define a critical point p_0 to be non-degenerate if it satisfies

(1.7) $$\frac{\partial^2 f}{\partial x^2}(p_0) \neq 0, \quad \frac{\partial^2 f}{\partial y^2}(p_0) \neq 0.$$

This is, in fact, a "bad definition," since after some coordinate changes, the condition (1.7) would no longer hold for the same f and p_0 in general. We want the concept of *non-degenerate critical points* or that of *degenerate critical points* to be independent of choice of coordinates. The following definition satisfies this requirement.

DEFINITION 1.4. (i) Suppose that $p_0 = (x_0, y_0)$ is a critical point of a function $z = f(x, y)$. We call the matrix

(1.8) $$\begin{pmatrix} \dfrac{\partial^2 f}{\partial x^2}(p_0) & \dfrac{\partial^2 f}{\partial x \partial y}(p_0) \\ \dfrac{\partial^2 f}{\partial y \partial x}(p_0) & \dfrac{\partial^2 f}{\partial y^2}(p_0) \end{pmatrix},$$

of second derivatives evaluated at p_0, the *Hessian* of the function $z = f(x, y)$ at a critical point p_0, and denote it by $H_f(p_0)$.

(ii) A critical point p_0 of a function $z = f(x, y)$ is *non-degenerate* if the determinant of the Hessian of f at p_0 is not zero; that is, p_0 is non-degenerate if we have the following:

(1.9) $$\det H_f(p_0) = \frac{\partial^2 f}{\partial x^2}(p_0)\frac{\partial^2 f}{\partial y^2}(p_0) - \left(\frac{\partial^2 f}{\partial x \partial y}(p_0)\right)^2 \neq 0.$$

On the other hand, if $\det H_f(p_0) = 0$, we say that p_0 is a *degenerate critical point*.

Notice that the matrix $H_f(p_0)$ is a symmetric matrix, since $\dfrac{\partial^2 f}{\partial x \partial y} = \dfrac{\partial^2 f}{\partial y \partial x}$.

EXAMPLE 1.5. Let us compute the Hessian for each of the three functions in Example 1.3 evaluated at the origin $\mathbf{0}$.

(i) For $z = x^2 + y^2$, the Hessian at the origin is

$$\begin{pmatrix} 2 & 0 \\ 0 & 2 \end{pmatrix}.$$

(ii) For $z = x^2 - y^2$, we have

$$\begin{pmatrix} 2 & 0 \\ 0 & -2 \end{pmatrix}.$$

(iii) For $-x^2 - y^2$, we have

$$\begin{pmatrix} -2 & 0 \\ 0 & -2 \end{pmatrix}.$$

The determinant of each of these matrices is not zero, and hence the origin $\mathbf{0}$ is a non-degenerate critical point for each of the three functions.

EXAMPLE 1.6. Consider the function $z = xy$. The origin $\mathbf{0}$ is its critical point. The Hessian at $\mathbf{0}$ is

$$\begin{pmatrix} 0 & 1 \\ 1 & 0 \end{pmatrix},$$

and its determinant is not zero; hence, the origin $\mathbf{0}$ is a non-degenerate critical point. In fact, the function $z = xy$ is obtained from $z = x^2 - y^2$ by a coordinate change.

EXAMPLE 1.7. The origin $\mathbf{0}$ is a critical point of the function $z = x^2 + y^3$, but the Hessian of this function at $\mathbf{0}$ is

$$\begin{pmatrix} 2 & 0 \\ 0 & 0 \end{pmatrix},$$

whose determinant is zero. Thus $\mathbf{0}$ is a degenerate critical point of $z = x^2 + y^3$.

How does the Hessian change through coordinate changes?

LEMMA 1.8. *Let p_0 be a critical point of a function $z = f(x, y)$. Denote by $H_f(p_0)$ the Hessian of f computed using the coordinates (x, y), and by $\mathcal{H}_f(p_0)$ the Hessian of the same f computed using different coordinates (X, Y). Then the following relation holds:*

$$(1.10) \qquad \mathcal{H}_f(p_0) = {}^t\!J(p_0) H_f(p_0) J(p_0),$$

where $J(p_0)$ is the Jacobian matrix for the above coordinate transformation, defined by

$$(1.11) \qquad J(p_0) = \begin{pmatrix} \dfrac{\partial x}{\partial X}(p_0) & \dfrac{\partial x}{\partial Y}(p_0) \\[2ex] \dfrac{\partial y}{\partial X}(p_0) & \dfrac{\partial y}{\partial Y}(p_0) \end{pmatrix},$$

and ${}^t\!J(p_0)$ is the transpose matrix of $J(p_0)$.

PROOF. The proof is by a simple calculation. We apply twice the well-known formula for the change of variables in partial derivatives:

$$\frac{\partial f}{\partial X} = \frac{\partial f}{\partial x}\frac{\partial x}{\partial X} + \frac{\partial f}{\partial y}\frac{\partial y}{\partial X},$$

$$\frac{\partial f}{\partial Y} = \frac{\partial f}{\partial x}\frac{\partial x}{\partial Y} + \frac{\partial f}{\partial y}\frac{\partial y}{\partial Y},$$

and express $\dfrac{\partial^2 f}{\partial X^2}$, $\dfrac{\partial^2 f}{\partial X \partial Y}$, and $\dfrac{\partial^2 f}{\partial Y^2}$ in terms of $\dfrac{\partial^2 f}{\partial x^2}$, $\dfrac{\partial^2 f}{\partial x \partial y}$, and $\dfrac{\partial^2 f}{\partial y^2}$. When we evaluate the values of these partial derivatives at p_0, we note that $\dfrac{\partial f}{\partial x}(p_0) = 0$ and $\dfrac{\partial f}{\partial y}(p_0) = 0$, since p_0 is a critical point of f. We leave the details of computation to the reader (in fact, this calculation is slightly complicated, so the reader may take this lemma for granted). $\qquad\square$

EXAMPLE 1.9. We rewrite the function $z = xy$ (Example 1.6) using the new coordinates

$$(1.12) \qquad \begin{cases} x = X - Y, \\ y = X + Y. \end{cases}$$

We obtain

$$xy = (X - Y)(X + Y) = X^2 - Y^2,$$

which is the second function in Example 1.3. The Hessians at the origin $\mathbf{0}$ (a critical point) of these functions with respect to the coordinates (x, y) and (X, Y) were

$$\begin{pmatrix} 0 & 1 \\ 1 & 0 \end{pmatrix}, \quad \begin{pmatrix} 2 & 0 \\ 0 & -2 \end{pmatrix},$$

respectively. The Jacobian matrix for the coordinate transformation (1.12) is

$$\begin{pmatrix} 1 & -1 \\ 1 & 1 \end{pmatrix},$$

and so the relation of Lemma 1.8

$$\begin{pmatrix} 2 & 0 \\ 0 & -2 \end{pmatrix} = \begin{pmatrix} 1 & 1 \\ -1 & 1 \end{pmatrix} \begin{pmatrix} 0 & 1 \\ 1 & 0 \end{pmatrix} \begin{pmatrix} 1 & -1 \\ 1 & 1 \end{pmatrix}$$

certainly holds.

The following corollary is a consequence of Lemma 1.8.

COROLLARY 1.10. *The property that p_0 is a non-degenerate critical point does not depend on choice of coordinates. The same is true for degenerate critical points.*

In fact, we have $\mathcal{H}_f(p_0) = {}^t\!J(p_0) H_f(p_0) J(p_0)$ by Lemma 1.8, and hence

$$(1.13) \qquad \det \mathcal{H}_f(p_0) = \det {}^t\!J(p_0) \det H_f(p_0) \det J(p_0),$$

by taking the determinants of the matrices on both sides. Noticing that the determinant of the Jacobian matrix satisfies

$$(1.14) \qquad \det J(p_0) \neq 0,$$

we conclude that $\det \mathcal{H}_f(p_0) \neq 0$ and $\det H_f(p_0) \neq 0$ are equivalent statements. This proves Corollary 1.10.

1.3. The Morse lemma

In this section we prove the following fact.

THEOREM 1.11 (The Morse lemma). *Let p_0 be a non-degenerate critical point of a function f of two variables. Then we can choose appropriate local coordinates (X, Y) in such a way that the function f expressed with respect to (X, Y) takes one of the following three standard forms:*

(i) $f = X^2 + Y^2 + c,$
(ii) $f = X^2 - Y^2 + c,$
(iii) $f = -X^2 - Y^2 + c,$

where c is a constant $(c = f(p_0))$ and p_0 is the origin $(p_0 = (0,0))$.

This theorem says that a function looks extremely simple near a non-degenerate critical point: for a function of two variables, a suitable coordinate change will make it one of the three simple functions we saw in Example 1.3.

None of these three standard functions has any other critical points near the origin, which is a non-degenerate critical point. Thus we obtain the following

COROLLARY 1.12. *A non-degenerate critical point of a function of two variables is isolated.*

In fact, this holds for functions of m variables, as well as those with two variables.

We now prove Theorem 1.11.

PROOF. Choose any local coordinate system (x, y) near the point p_0. We may assume that the point p_0 is the origin $(0,0)$ in these coordinates. Throughout the following discussion we may assume for simplicity that $f(p_0) = 0$. Further we will show that we may assume

(1.15) $\dfrac{\partial^2 f}{\partial x^2}(p_0) \neq 0.$

Of course there is nothing to prove if we already have $\dfrac{\partial^2 f}{\partial x^2}(p_0) \neq 0$.

Also if $\dfrac{\partial^2 f}{\partial y^2}(p_0) \neq 0$, then even if $\dfrac{\partial^2 f}{\partial x^2}(p_0) = 0$, by interchanging the x-axis and the y-axis, we may assume that the condition (1.15) is valid. So we consider the case where $\dfrac{\partial^2 f}{\partial x^2}(p_0) = 0$ and $\dfrac{\partial^2 f}{\partial y^2}(p_0) = 0$.

In this case the Hessian H_f with respect to (x, y) (at the point p_0) is as follows:

(1.16) $H_f = \begin{pmatrix} 0 & a \\ a & 0 \end{pmatrix}, \quad a \neq 0.$

Here we specify that $a \neq 0$, because p_0 is a non-degenerate critical point. Introduce a new local coordinate system (X, Y) by

$$(1.17) \qquad x = X - Y, \quad y = X + Y.$$

Then the Jacobian J for the change of coordinates from (X, Y) to (x, y) is

$$(1.18) \qquad J = \begin{pmatrix} 1 & -1 \\ 1 & 1 \end{pmatrix},$$

so that the Hessian $\mathcal{H}_f$ with respect to (X, Y) is

$$(1.19) \qquad \mathcal{H}_f = {}^t\! J H_f J = \begin{pmatrix} 2a & 0 \\ 0 & -2a \end{pmatrix}$$

by Lemma 1.8. This equality shows that

$$(1.20) \qquad \frac{\partial^2 f}{\partial X^2}(p_0) = 2a \neq 0, \quad \frac{\partial^2 f}{\partial Y^2}(p_0) = -2a \neq 0.$$

Using the old notation (x, y) for (X, Y), we see that f certainly satisfies the condition (1.15). Thus in the following we may proceed with the assumption (1.15).

We recall the following fact from the calculus of several variables. Suppose we have a function $z = f(x, y)$ defined near the origin with $f(0, 0) = 0$. Then there are functions $g(x, y)$ and $h(x, y)$ such that we can write

$$(1.21) \qquad f(x, y) = xg(x, y) + yh(x, y)$$

in some neighborhood of the origin $(0, 0)$, and such that

$$(1.22) \qquad \frac{\partial f}{\partial x}(0, 0) = g(0, 0), \quad \frac{\partial f}{\partial y}(0, 0) = h(0, 0).$$

We will prove this fact first.

We assume for simplicity that $z = f(x, y)$ is defined in the entire xy-plane. Choose an arbitrary point (x, y), which will stay fixed. Consider a function $f(tx, ty)$ with parameter t. If we differentiate this function with respect to t and then integrate, we obtain the original one back. In particular, if we look at its definite integral from 0 to 1 together with the condition $f(0,0) = 0$, we have

$$
\begin{aligned}
f(x,y) &= \int_0^1 \frac{df(tx,ty)}{dt}\, dt \\
&= \int_0^1 \left\{ x\frac{\partial f}{\partial x}(tx,ty) + y\frac{\partial f}{\partial y}(tx,ty) \right\} dt \\
&= xg(x,y) + yh(x,y).
\end{aligned}
$$

(1.23)

In the middle of the above equalities, we used the chain rule for composite functions (cf. [19]). The term $\dfrac{\partial f}{\partial x}(tx,ty)$ may look somewhat confusing, but it is nothing but the value of the derivative $\dfrac{\partial f}{\partial x}$ of the function $f(x,y)$ evaluated at the point (tx,ty). This is also the case for $\dfrac{\partial f}{\partial y}(tx,ty)$. In the last line of the equality (1.23), we set

(1.24) $\quad g(x,y) = \displaystyle\int_0^1 \frac{\partial f}{\partial x}(tx,ty)\, dt, \quad h(x,y) = \int_0^1 \frac{\partial f}{\partial y}(tx,ty)\, dt.$

Thus we have shown the equality (1.21). Furthermore, substituting $(x,y) = (0,0)$ in (1.24), we establish the equality (1.22).

The above fact is rather fundamental in the calculus of several variables, but we have presented a proof here for the reader's convenience.

Now in our case, since we assume that the origin $p_0 = (0,0)$ is a critical point of the function f, we have

(1.25) $\quad g(0,0) = \dfrac{\partial f}{\partial x}(0,0) = 0, \quad h(0,0) = \dfrac{\partial f}{\partial y}(0,0) = 0.$

Hence we can apply the fundamental fact from calculus proved above again to our functions $g(x,y)$ and $h(x,y)$, and with suitable functions h_{11}, h_{12}, h_{21} and h_{22} we can write

(1.26) $\quad\quad\quad\quad g(x,y) = xh_{11}(x,y) + yh_{12}(x,y)$

and

(1.27) $\quad\quad\quad\quad h(x,y) = xh_{21}(x,y) + yh_{22}(x,y).$

We combine the equalities (1.26) and (1.27) with the equality (1.21) to obtain

(1.28) $\quad\quad\quad f(x,y) = x^2 h_{11} + xy(h_{12} + h_{21}) + y^2 h_{22}.$

The substitutions $H_{11} = h_{11}$, $H_{12} = (h_{12} + h_{21})/2$, and $H_{22} = h_{22}$ change (1.28) to

(1.29) $f(x,y) = x^2 H_{11} + 2xy H_{12} + y^2 H_{22}$,

which has a simpler form. From this equality a simple calculation gives us

(1.30)
$$\begin{cases} \dfrac{\partial^2 f}{\partial x^2}(0,0) = 2H_{11}(0,0), \\[2mm] \dfrac{\partial^2 f}{\partial x \partial y}(0,0) = \dfrac{\partial^2 f}{\partial y \partial x}(0,0) = 2H_{12}(0,0), \\[2mm] \dfrac{\partial^2 f}{\partial y^2}(0,0) = 2H_{22}(0,0). \end{cases}$$

Keep in mind that the left-hand side of the first equality is assumed to be not zero. Hence $H_{11}(0,0) \neq 0$, and since H_{11} is continuous, we see that

(1.31) $H_{11}(x,y)$ is not zero in some neighborhood of $(0,0)$.

We now define a new x-coordinate X near this neighborhood of the origin $(0,0)$ by

(1.32) $X = \sqrt{|H_{11}|} \left(x + \dfrac{H_{12}}{H_{11}} y \right).$

We keep the y-coordinate as it is. The Jacobian between (x,y) and (X,y) evaluated at the origin is not zero, so (X,y) is certainly a local coordinate system for some neighborhood of the origin $(0,0)$. We square X:

(1.33)
$$X^2 = |H_{11}| \left(x^2 + 2\dfrac{H_{12}}{H_{11}} xy + \dfrac{H_{12}^2}{H_{11}^2} y^2 \right)$$
$$= \begin{cases} H_{11}x^2 + 2H_{12}xy + \dfrac{H_{12}^2}{H_{11}} y^2 & (H_{11} > 0), \\[3mm] -H_{11}x^2 - 2H_{12}xy - \dfrac{H_{12}^2}{H_{11}} y^2 & (H_{11} < 0). \end{cases}$$

If we compare (1.33) with (1.29), we see that for $H_{11} > 0$

(1.34) $f = X^2 + \left(H_{22} - \dfrac{H_{12}^2}{H_{11}} \right) y^2,$

and for $H_{11} < 0$

(1.35) $f = -X^2 + \left(H_{22} - \dfrac{H_{12}^2}{H_{11}} \right) y^2.$

From equality (1.30) we obtain

$$(1.36) \qquad H_{11}(0,0)H_{22}(0,0) - H_{12}^2(0,0) = \frac{1}{4} \det H_f \neq 0,$$

where $\det H_f \neq 0$ because of the assumption that the origin p_0 is a non-degenerate critical point of f. If we choose a new y-coordinate Y near the origin $p_0 = (0,0)$ by

$$(1.37) \qquad Y = \sqrt{\left| \frac{H_{11}H_{22} - H_{12}^2}{H_{11}} \right|} \; y,$$

and rewrite the equalities (1.34) and (1.35), f has the following expression in the local coordinates (X, Y):

$$(1.38) \qquad f = \begin{cases} X^2 + Y^2 & (H_{11} > 0, \; K > 0), \\ X^2 - Y^2 & (H_{11} > 0, \; K < 0), \\ -X^2 + Y^2 & (H_{11} < 0, \; K < 0), \\ -X^2 - Y^2 & (H_{11} < 0, \; K > 0), \end{cases}$$

where we denote $H_{11}H_{22} - H_{12}^2$ by K for simplicity. If we interchange X and Y through a "90° rotation", we see that $f = X^2 - Y^2$ and $f = -X^2 + Y^2$ are essentially the same standard form. This concludes the proof of Theorem 1.11. $\qquad\qquad\qquad\qquad\qquad\qquad\qquad$ □

DEFINITION 1.13 (Index of a non-degenerate critical point). Let p_0 be a non-degenerate critical point of a function f of two variables. We choose a suitable coordinate system (x, y) in some neighborhood of the point p_0 so that the function f has a standard form given by Theorem 1.11. Then we define the *index* of the non-degenerate critical point p_0 of f to be 0, 1 and 2, respectively for $f = x^2 + y^2 + c$, $f = x^2 - y^2 + c$ and $f = -x^2 - y^2 + c$. In other words, the number of minus signs in the standard form is the index of p_0.

We see immediately from the respective graphs (Figure 1.4) of the functions $z = x^2 + y^2$, $z = x^2 - y^2$ and $z = -x^2 - y^2$ that if the point p_0 has index 0, then f takes a minimum value at p_0. If the index of p_0 is 1, then in some neighborhood of p_0, f may take values strictly larger than $f(p_0)$ or it may take values strictly smaller than $f(p_0)$. If the index of p_0 is 2, then f takes a maximum value at p_0. Thus the index of a non-degenerate critical point p_0 is determined by the behavior of f near p_0.

We might as well prove that the index is well-defined as follows. Recall that in Example 1.5 we computed that the Hessians of the functions $z = x^2 + y^2$, $z = x^2 - y^2$, and $z = -x^2 - y^2$ are

$$\begin{pmatrix} 2 & 0 \\ 0 & 2 \end{pmatrix}, \quad \begin{pmatrix} 2 & 0 \\ 0 & -2 \end{pmatrix}, \quad \text{and} \quad \begin{pmatrix} -2 & 0 \\ 0 & -2 \end{pmatrix}$$

respectively.

We may think of these matrices as the diagonalized Hessian $H_f(p_0)$ of f. In other words, there is a matrix J such that the matrix ${}^t\!J H_f(p_0) J$ is a diagonal matrix (one of the above), that is, a matrix with non-zero numbers only on diagonal entries. According to Sylvester's law in linear algebra (see for example [20]), when we diagonalize a symmetric matrix such as our $H_f(p_0)$, the number of minus signs appearing in the diagonal depends on $H_f(p_0)$ and not on the way we diagonalize it. This proves that the index is well-defined.

In Chapter 2 we will prove an analogue of Theorem 1.11 in the m-dimensional space in general, and will define the index of a non-degenerate critical point p_0. Sylvester's law will guarantee the well-definedness of the index in that case as well.

1.4. Morse functions on surfaces

In the previous sections we limited ourselves to "local" investigation of critical points in their neighborhoods. We now turn to a "global" investigation which involves the shape of a space as a whole. In this section we consider two–dimensional spaces; that is, *surfaces*.

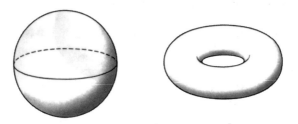

FIGURE 1.5. A sphere and a torus

Some examples of *closed surfaces* are depicted in Figures 1.5 and 1.6: a sphere and a torus in Figure 1.5, and closed surfaces of genus two and three in Figure 1.6. By the *genus* of a closed surface we mean the number of "holes" in it. The genus of a torus is one and that of a sphere is zero. We consider a closed surface of genus g for any natural number g. If one thinks of a torus as a "float," then one might think

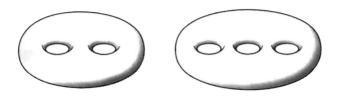

FIGURE 1.6. Closed surfaces of genus 2 and genus 3

of a surface of genus two as a "float for two persons." Similarly we may think of a surface of genus g as a "float for g persons." (This expression is due to M. Kuga.)

We denote the sphere by S^2. The superscript 2 represents the dimension of the sphere. We denote a torus by T^2. We often denote by Σ_g the closed surface of genus g, and in this case Σ_0 and Σ_1 are nothing but a sphere S^2 and a torus T^2, respectively.

Let M be a surface. We call a map

$$f : M \to \mathbb{R},$$

which assigns a real number to each point p of M, a function on M. Here $\mathbb{R}$ denotes the set of all real numbers.

Notice that a surface is curved, so that local coordinates on it are also curved in general (cf. Figure 1.7).

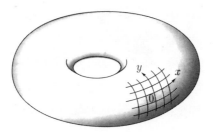

FIGURE 1.7. A local coordinate system on a surface

We say that a function $f : M \to \mathbb{R}$ defined on a surface M is of class C^∞ (or smooth) if it is of class C^∞ with respect to any smooth local coordinates at each point of M.

From now on we will consider only smooth (C^∞) surfaces and smooth (C^∞) functions defined on them.

The concept of a "critical point" we saw in the previous section carries over to a function $f : M \to \mathbb{R}$ defined on a surface M with the aid of local coordinates. More precisely, we say that a point p_0 of a surface M is a *critical point* of a function $f : M \to \mathbb{R}$ if

$$(1.39) \qquad \frac{\partial f}{\partial x}(p_0) = 0, \quad \frac{\partial f}{\partial y}(p_0) = 0$$

with respect to local coordinates in some neighborhood of p_0. We saw in the first section that non-degenerate critical points are stable and have some convenient properties in contrast to degenerate critical points. Therefore the functions on a surface with only non-degenerate critical points would be nice ones. Based on this consideration, we define

DEFINITION 1.14 (Morse functions). Suppose that every critical point of a function $f : M \to \mathbb{R}$ on M is non-degenerate. Then we say that f is a *Morse function*.

Let us look at some examples of Morse functions.

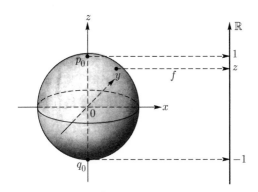

FIGURE 1.8. A height function $f : S^2 \to \mathbb{R}$

EXAMPLE 1.15 (The height function on the sphere). We consider the unit sphere S^2 with the orthogonal coordinates (x, y, z) in three-dimensional space $\mathbb{R}^3$; that is, S^2 is defined by the equation

$$(1.40) \qquad x^2 + y^2 + z^2 = 1.$$

Let $f : S^2 \to \mathbb{R}$ be a function on S^2 which assigns to each point $p = (x, y, z)$ on S^2 its third coordinate z. One might say that f is the "height function." Then f is a Morse function (Figure 1.8).

In fact f has two critical points; the north pole $p_0 = (0, 0, 1)$ and the south pole $q_0 = (0, 0, -1)$. One easily sees that f has no other critical points. In order to show that f is a Morse function, it is enough to prove that p_0 and q_0 are both non-degenerate. To do this, one computes the Hessian of f with respect to the coordinate system (x, y).

As we see in the above example, there is a Morse function on a sphere with exactly two critical points, both of which are non-degenerate. In fact, we can show the converse.

THEOREM 1.16. *Let M be a closed surface. Suppose that there exists a Morse function $f : M \to \mathbb{R}$ with exactly two non-degenerate critical points (and no other critical points). Then M is diffeomorphic to the sphere S^2.*

This theorem is a simple example describing Morse theory. Before proving the theorem, we must define a "diffeomorphism."

We start with the concept of a "homeomorphism." Suppose that there is a one-to-one and "onto" map

$$h : X \to Y$$

between two (topological) spaces X and Y; that is, every point of X corresponds to a point of Y and every point of y to a point of X in a one-to-one fashion. Then we can define the inverse map

$$h^{-1} : Y \to X.$$

If both of the maps $h : X \to Y$ and $h^{-1} : Y \to X$ are continuous, we say that $h : X \to Y$ is a *homeomorphism*. Two (topological) spaces X and Y are *homeomorphic* if there is a homeomorphism between them. If X and Y are homeomorphic, then in topology, we consider that they have "the same shape."

DEFINITION 1.17. A homeomorphism

$$h : M \to N$$

between surfaces M and N is a *diffeomorphism* if the maps $h : M \to N$ and $h^{-1} : N \to M$ are both of class C^∞. Two surfaces M and N are called *diffeomorphic* if there is a diffeomorphism $h : M \to N$ between them.

Two diffeomorphic surfaces M and N are considered to have an identical shape with their smooth structures taken into consideration. In differential topology, where the objects of study are smooth

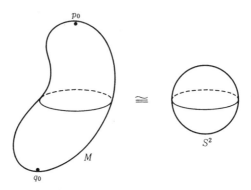

FIGURE 1.9. M is diffeomorphic to S^2

geometrical figures, one considers two diffeomorphic spaces as having "the same shape."

Let us prove Theorem 1.16 (cf. Figure 1.9).

PROOF. Notice that a closed surface M is compact (in fact, we define a closed surface as a "two-dimensional compact manifold without boundary"). We will review the concept of compactness in Chapter 2. In our proof, we take for granted the following maximum–value theorem, which is one of the properties that compact spaces enjoy.

THEOREM 1.18 (Maximum–value theorem). *Let $f : X \to \mathbb{R}$ be a continuous function defined on a compact space X. Then f takes the maximum value at some point p_0 and the minimum value at some point q_0.*

The maximum–value theorem is well-known, and so we omit its proof (see for example [19]).

By the maximum–value theorem, the Morse function $f : M \to \mathbb{R}$ takes the maximum value at some point p_0 and the minimum value at some other point q_0 in M. The points p_0 and q_0 are critical points of f. Moreover, since f is a Morse function, p_0 and q_0 are non-degenerate critical points.

According to the Morse lemma (Theorem 1.11), one can express f in a standard form with a suitable coordinate system (x, y) about p_0 and another coordinate system (X, Y) about q_0.

The index of p_0 is 2 since f takes the maximum value at this point, and the index of q_0 is 0 because f takes the minimum value

FIGURE 1.10. Left: the graph of f near p_0; Right: the graph of f near q_0

there. Thus we have the following expression for f in these coordinate systems:

$$(1.41) \qquad f = \begin{cases} -x^2 - y^2 + A & \text{(near } p_0), \\ X^2 + Y^2 + a & \text{(near } q_0). \end{cases}$$

Here A and a are the maximum and minimum values of f respectively. Let ε be a small enough positive number, and denote by $D(p_0)$ the set of points in a neighborhood of p_0 with

$$(1.42) \qquad A - \varepsilon \le f(p) \le A.$$

The upside-down bowl in the left of Figure 1.10 describes the set $D(p_0)$ which, in the local coordinates (x, y), corresponds to the inequality

$$(1.43) \qquad x^2 + y^2 \le \varepsilon,$$

according to the equality (1.41). In other words, $D(p_0)$ is diffeomorphic to the 2-*disk* (2-dimensional disk) defined by the the the inequality (1.43). Similarly the set $D(q_0)$ of points in a neighborhood of q_0, satisfying

$$(1.44) \qquad a \le f(p) \le a + \varepsilon,$$

corresponds to the (right-side-up) bowl in the right of Figure 1.10. In terms of the local coordinates (X, Y), $D(q_0)$ has the expression

$$(1.45) \qquad X^2 + Y^2 \le \varepsilon,$$

and so it is also diffeomorphic to the 2-disk.

We now remove the interiors of $D(p_0)$ and $D(q_0)$ from the surface M, and denote by M_0 the resulting "surface with boundary" (Figure 1.11). The boundary of M_0 (above and below) consists of the boundary circles $C(p_0)$ and $C(q_0)$ of $D(p_0)$ and $D(q_0)$ respectively.

In general, we denote the boundary by

$$(1.46) \qquad \partial M_0$$

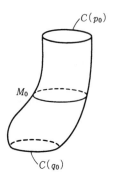

FIGURE 1.11. The surface M_0

for a surface M_0 with boundary. In our case, the boundary of M_0 consists of two circles $C(p_0)$ and $C(q_0)$, so that we write

$$(1.47) \qquad \partial M_0 = C(p_0) \cup C(q_0).$$

We denote by

$$(1.48) \qquad \mathrm{int}(M_0)$$

the *interior* of M_0, which is what is left when the boundary is removed from M_0. By definition, $\mathrm{int}(M_0) = M_0 - \partial M_0$.

It is evident from the definition of $D(p_0)$ and $D(q_0)$ (the inequalities (1.42) and (1.44)) that the restriction of the function f to M_0, $f : M_0 \to \mathbb{R}$, takes the constant values $A - \varepsilon$ and $a + \varepsilon$ on the boundary circles $C(p_0)$ and $C(q_0)$ respectively.

Recall that in Theorem 1.16, it was assumed that the Morse function has exactly two critical points p_0 and q_0. Thus, having removed (the interior of) $D(p_0)$ and $D(q_0)$, we no longer have any critical points for $f : M_0 \to \mathbb{R}$. Under this circumstance we have the following fact.

LEMMA 1.19. *Let $f : M_0 \to \mathbb{R}$ be a C^∞-function which takes constant values on the boundary circles $C(p_0)$ and $C(q_0)$. We further assume that f has no critical points on M_0. Then M_0 is diffeomorphic to the direct product of one of the boundary circles, say $C(q_0)$, and the unit interval, $C(q_0) \times [0, 1]$.*

We will prove Lemma 1.19 in the next chapter (Theorem 2.31) in a more general setting.

Since the boundary $C(q_0)$ is diffeomorphic to the unit circle (denoted by S^1), Lemma 1.19 implies that M_0 is diffeomorphic to the direct product

$$S^1 \times [0, 1].$$

In general, a surface diffeomorphic to the direct product $S^1 \times [0, 1]$ is called an *annulus*. For example, if we remove from a disk Δ the interior of a small concentric disk Δ_0, the resulting suface A is an annulus.

By Lemma 1.19, M_0 is an annulus. Set

$$N_0 = M_0 \cup D(q_0);$$

that is, N_0 is the union of M_0 and the (upright) bowl $D(q_0)$. More precisely, one obtains N_0 by attaching the disk $D(q_0)$ to M_0 along the boundary of $D(q_0)$, and therefore N_0 is also diffeomorphic to the disk.

We paste $D(p_0)$ along the boundary of N_0 (they share the common boundary $C(p_0)$) to recover M. This shows that M is a closed surface diffeomorphic to the sphere S^2, and the proof of Theorem 1.16 is complete.   □

Strictly speaking, we need the following lemma (Exercise 1.1) to show that the closed surface we obtain from two disks by pasting their boundary circles together is diffeomorphic to the sphere S^2.

LEMMA 1.20. *Let*

(1.49) $k : \partial D_0 \to \partial D_1$

be a diffeomorphism between the respective boundaries of two disks D_0 and D_1. Then we can extend k to a diffeomorphism

(1.50) $K : D_0 \to D_1$

of the disks.

See Exercise 1.3 and its solution for the proof of Lemma 1.20. We emphasize here that this lemma is not a "trivial fact". In fact, it holds for disks of dimension less than or equal to six, but is not always true for disks of dimension higher than six.

Before closing this section we prove the following elementary fact.

LEMMA 1.21. *A Morse function $f : M \to \mathbb{R}$ defined on a closed surface M has only a finite number of critical points.*

PROOF. We derive a contradiction assuming that a Morse function $f : M \to \mathbb{R}$ on a closed surface M has infinitely many critical points

$$p_1, p_2, p_3, \dots .$$

Since the closed surface M is compact, there is a convergent subsequence $p_{n(1)}, p_{n(2)}, \dots$ of this sequence. (See for example [19].) Let p_0 be its limit point. Consider a local coordinate system (x, y) defined on a neighborhood U of p_0. Since the above subsequence $\{p_{n(i)}\}_{i=1}^{\infty}$ converges to the point p_0, by choosing a further subsequence if necessary, we may assume that the subsequence $\{p_{n(i)}\}_{i=1}^{\infty}$ is contained in the neighborhood U of p_0.

Since f is of class C^{∞}, its partial derivatives $\dfrac{\partial f}{\partial x}(p)$ and $\dfrac{\partial f}{\partial y}(p)$ depend smoothly on the point p. The derivatives $\dfrac{\partial f}{\partial x}(p)$ and $\dfrac{\partial f}{\partial y}(p)$ take the value zero on the sequence $p_{n(1)}, p_{n(2)}, \dots$ of critical points. Hence these derivatives take the value zero at p_0 as well, since p_0 is the limit point of $p_{n(1)}, p_{n(2)}, \dots .$ Thus p_0 is a critical point of the function f. All the critical points of the Morse function f are non-degenerate, and they are isolated by Corollary 1.12. However, the sequence $\{p_{n(i)}\}_{i=1}^{\infty}$ consisting of critical points converges to a critical point p_0, which is a contradiction. This completes the proof of Lemma 1.21.   □

1.5. Handle decomposition

In Theorem 1.16 in the preceding section, we saw that a Morse function on a surface determines the shape of the surface, for a special case. Morse theory (especially Morse theory for finite-dimensional spaces) is a systematic study of this phenomenon. We adopt a powerful tool called "handle decompositions" for such a study. In this section, we use surfaces to explain handle decompositions.

We start with a Morse function

$$f : M \to \mathbb{R}$$

defined on a closed surface.

In what follows, we assume that the closed surface M is *connected*. For a surface M (or more generally for a manifold M) this assumption is equivalent to assuming that M is *arcwise connected*, i.e., any two points p and q on M are joined by an arc drawn on M.

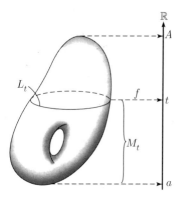

FIGURE 1.12. A subsurface M_t and a level curve L_t

Given a Morse function $f : M \to \mathbb{R}$, we denote by M_t the "subsurface" of M consisting of all points at which f takes values less than or equal to the real number t; that is, we set

$$(1.51) \qquad M_t = \{\, p \in M \mid f(p) \leq t \,\}.$$

Denote by L_t the set of points where the value of f is exactly t. We call L_t the level curve at $f = t$. The subsurface M_t is everything below the level curve L_t. The level curve L_t is the boundary of the subsurface M_t (cf. Figure 1.12).

The function $f : M \to \mathbb{R}$ takes maximum value A and minimum value a. Since there is no point p with $f(p) < a$, we have

$$M_t = \emptyset$$

for $t < a$. We also note that $f(p)$ is always less than or equal to the maximum value A at any point p of M, so that we have $f(p) \leq t$ at any point p of M if $A \leq t$; that is, if $A \leq t$ then

$$M_t = M.$$

Thus, as the parameter t starts from a value less than a and increases until it passes A, the subsurface M_t starts from the empty set and changes until it becomes the entire M. The fundamental idea of Morse theory is to trace this change of shapes of M_t.

We may also think of $f : M \to \mathbb{R}$ as a height function and think of sinking M gradually in the water as follows. The parameter t indicates the water level, and M_t is the portion of the surface under

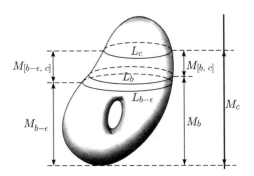

FIGURE 1.13. There is no critical point between L_b and L_c

water when the water level is t. As the water level increases, the shape of M_t under water changes. Morse theory studies the change of the shape of M_t under water.

DEFINITION 1.22 (Critical values). We say that a real number c_0 is a *critical value* of f if f takes the value c_0 at some critical point p_0: $f(c_0) = p_0$.

LEMMA 1.23. *Let $b < c$ be real numbers such that f has no critical values in the interval $[b, c]$. Then M_b and M_c are diffeomorphic (cf. Figure 1.13).*

PROOF. Since the minimum and maximum values a and A, respectively, are critical values for f, the interval $[b, c]$ does not contain a or A. We may thus prove Lemma 1.23 under the assumption that

$$a < b < c < A.$$

Denote by $M_{[b,c]}$ the portion between the level curves L_b and L_c:

$$(1.52) \qquad M_{[b,c]} = \{ p \in M \mid b \le f(p) \le c \}.$$

Clearly we have

$$(1.53) \qquad M_b \cup M_{[b,c]} = M_c.$$

By assumption $M_{[b,c]}$ contains no critical points of f.

By Lemma 1.21, $f : M \to \mathbb{R}$ has only a finite number of critical points. We may therefore assume that f has no critical points in $M_{[b-\varepsilon,c]}$ (which we obtain by adding a short skirt at the bottom of $M[b, c]$), for a small enough positive number ε.

According to Theorem 2.31, which we will prove in the next chapter, $M_{[b-\varepsilon,c]}$ is diffeomorphic to the product $L_{b-\varepsilon} \times [0,1]$ in this case. We also note that $M_{[b-\varepsilon,b]} \subset M_{[b-\varepsilon,c]}$ so that f has no critical points in $M_{[b-\varepsilon,b]}$, and hence $M_{[b-\varepsilon,b]}$ is also diffeomorphic to the product $L_{b-\varepsilon} \times [0,1]$, by the same theorem. Thus we have a diffeomorphism

$$h : M_{[b-\varepsilon,b]} \to M_{[b-\varepsilon,c]},$$

where we may assume that the restriction of h to the level curve $L_{b-\varepsilon}$ is the identity map. We now "glue" h and the identity map

$$\mathrm{id} : M_{b-\varepsilon} \to M_{b-\varepsilon}$$

along the boundary level curve $L_{b-\varepsilon}$ to define a diffeomorphism

$$H = \mathrm{id} \cup h : M_{b-\varepsilon} \cup M_{[b-\varepsilon,b]} \to M_{b-\varepsilon} \cup M_{[b-\varepsilon,c]}$$

("Gluing" diffeomorphisms has subtle problems. See Theorem 2.8 and the explanations thereafter in the next chapter).

Noting that $M_{b-\varepsilon} \cup M_{[b-\varepsilon,b]} = M_b$ and $M_{b-\varepsilon} \cup M_{[b-\varepsilon,c]} = M_c$, we indeed obtain the diffeomorphism

$$H : M_b \to M_c.$$

This completes the proof of Lemma 1.23.

The main idea of the proof is that if we keep stretching the narrow annulus $M_{[b-\varepsilon,b]}$ in M_b, the subsurface M_b will eventually cover M_c.
□

According to Lemma 1.23, we know that the shape of M_t stays unchanged as the parameter t goes through non-critical values of f, and thus what is important in keeping track of the M_t's is the change of M_t as t crosses a critical value c_0 of f.

If c_0 is a critical value, there exists at least one critical point p_0 with

$$f(p_0) = c_0.$$

For simplicity let us assume that there is only one critical point p_0 with the critical value c_0. We may then assume that for a small enough positive number ε, the only critical point of f in $M_{[c_0-\varepsilon,c_0+\varepsilon]}$ is p_0 ($M_{[c_0-\varepsilon,c_0+\varepsilon]}$ is the region between the level curves $L_{c_0-\varepsilon}$ and $L_{c_0+\varepsilon}$).

We investigate the relation between $M_{c_0-\varepsilon}$ and $M_{c_0+\varepsilon}$ as t crosses the critical value c_0.

(a) The case when the index of p_0 is zero

We choose a coordinate sytem (x, y) about p_0 and write f locally in standard form

(1.54) $$f = x^2 + y^2 + c_0,$$

using Theorem 1.11.

If c_0 is the minimum value of f, then $M_{c_0 - \varepsilon} = \emptyset$. Since $M_{c_0 + \varepsilon}$ is defined by

(1.55)
$$\begin{aligned}
M_{c_0 + \varepsilon} &= \{\, p \in M \mid f(p) \le c_0 + \varepsilon \,\} \\
&= \{\, (x, y) \mid x^2 + y^2 \le \varepsilon \,\},
\end{aligned}$$

which is a bowl diffeomorphic to the 2-disk D^2, we can describe the above change as follows: M_t is an empty set for t less than the minimum value c_0, but as soon as t passes the minimum value c_0, a disk pops out and M_t becomes diffeomorphic to the disk. If the critical value c_0 is not the minimum value, then $M_{c_0 - \varepsilon}$ is not empty. In this case, as t passes c_0 a disk will pop out and $M_{c_0 + \varepsilon}$ becomes diffeomorphic to the disjoint union of $M_{c_0 - \varepsilon}$ and the 2-disk D^2 (see Figure 1.14):

(1.56) $$M_{c_0 + \varepsilon} \cong M_{c_0 - \varepsilon} \sqcup D^2,$$

where the symbol $\cong$ means that the two sides of $\cong$ are diffeomorphic. Also we use $\sqcup$ to denote the *disjoint union*, which is the union of sets that do not intersect each other.

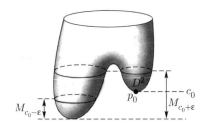

FIGURE 1.14. The case when the index of p_0 is zero

(b) The case when the index of p_0 is one

By Theorem 1.11, with a suitable coordinate system (x, y), f has the standard form

$$f = -x^2 + y^2 + c_0,$$

where we interchanged x and y in order to be consistent with the general theory in the next chapter. The graph near the critical point p_0 resembles a mountain pass, as shown in Figure 1.15. The curve corresponding to $y = 0$ on the graph goes downhill from p_0 and the curve $x = 0$ orthogonal to it goes uphill from p_0. The statement that the critical point p_0 has index 1, therefore, is equivalent to saying that the downward direction from p_0 is one-dimensional.

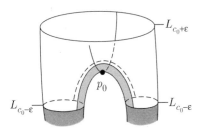

FIGURE 1.15. A graph near a critical point of index one

We add a small width to the downward curve in the graph at p_0 to make a path going over the mountain pass at p_0, which might also look like a bridge connecting the edges in $L_{c_0-\varepsilon}$ of $M_{c_0-\varepsilon}$. In Figure 1.16 we give a top view of the graph in Figure 1.15.

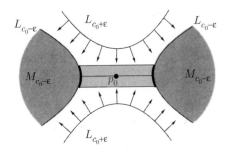

FIGURE 1.16. A top view of the graph

In Figure 1.16, the heavily shaded area defined by

$$-x^2 + y^2 \leq -\varepsilon$$

correponds to $M_{c_0-\varepsilon}$, and the lightly shaded area corresponds to the bridge on $M_{c_0-\varepsilon}$. This bridge is almost rectangular (diffeomorphic to a rectangle) although the right and left edges are slightly bent.

We call the interval $[0, 1]$ a 1-dimensional disk, or simply a 1-disk, and we often denote it by D^1. The end points of D^1 denoted by ∂D^1 form the boundary of D^1. Thus ∂D^1 consists of two points.

The lightly shaded bridge (a rectangle) in Figure 1.16 is diffeomorphic to the product $D^1 \times D^1$, and the intersection of the bridge with $M_{c_0-\varepsilon}$ (shown by thick lines in the figure) corresponds to the pair of parallel edges $\partial D^1 \times D^1$ (see Figure 1.17).

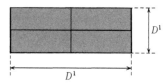

FIGURE 1.17. A 1-handle $D^1 \times D^1$ and the edges along which it is attached to $M_{c_0-\varepsilon}$

We say that the rectangle $D^1 \times D^1$ is a 1-*handle* attached to $M_{c_0-\varepsilon}$. The name, 1-handle, comes from the fact that it corresponds to a critical point of index one. We will get a better feel for the name when we encounter 3-dimensional handles in Chapter 3.

The manifold $M_{c_0+\varepsilon}$, right after the parameter t passes through the critical value c_0, corresponds in local coordinates to the portion

$$-x^2 + y^2 + c_0 \leq \varepsilon.$$

We compare $M_{c_0+\varepsilon}$ and $M_{c_0-\varepsilon}$ with a one-handle $D^1 \times D^1$ attached (Figure 1.16). Then we see that we can shrink $M_{c_0+\varepsilon}$ along the arrows as indicated in the figure until it coincides with $M_{c_0-\varepsilon} \cup D^1 \times D^1$. Hence we have

(1.57) $$M_{c_0+\varepsilon} \cong M_{c_0-\varepsilon} \cup D^1 \times D^1.$$

Thus for a critical point p_0 of index one, the change of M_t as the parameter t increases from $c_0 - \varepsilon$ to $c_0 + \varepsilon$ can be described as attaching a 1-handle to $M_{c_0-\varepsilon}$.

Strictly speaking, the surface $M_{c_0-\varepsilon} \cup D^1 \times D^1$ on the right-hand side of the diffeomorphism (1.57) is hardly a smooth surface, as the boundary of $M_{c_0-\varepsilon} \cup D^1 \times D^1$ bends by 90-degree angles at the four corners where the 1-handle $D^1 \times D^1$ is attached. We can, however, smooth out these corners by a well-known "smoothing" technique and can assume the right-hand side of the diffeomorphism in (1.57) to be a smooth surface. (See for example [12] or Remark 1.3.3 in [3].) Sometimes we think of $\cong$ in (1.57) as a homeomorphism, and whenever we want the right-hand side to be a smooth surface we simply consider $M_{c_0+\varepsilon}$ on the left-hand side instead. In either case, we continue our discussion assuming that $M_{c_0-\varepsilon} \cup D^1 \times D^1$ is smooth.

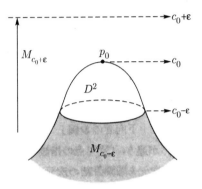

FIGURE 1.18. A critical point of index 2

(c) The case when the index of p_0 is two

In a local coordinate system (x, y) about p_0, by Theorem 1.11, f has the standard form

$$f = -x^2 - y^2 + c_0.$$

In the local coordinates (x, y) we can write

(1.58)
$$\begin{aligned} M_{c_0-\varepsilon} &= \{\, p \in M \mid f(p) \leq c_0 - \varepsilon \,\} \\ &= \{\, (x,y) \mid x^2 + y^2 \geq \varepsilon \,\}, \end{aligned}$$

and hence $M_{c_0-\varepsilon}$ is outside the disk of radius $\sqrt{\varepsilon}$ (the boundary circle is included). In Figure 1.18, we drew this disk of radius $\sqrt{\varepsilon}$ as an upside-down bowl, and thus we get $M_{c_0+\varepsilon}$ by capping $M_{c_0-\varepsilon}$ with this bowl along its boundary:

(1.59)
$$M_{c_0+\varepsilon} = M_{c_0-\varepsilon} \cup D^2.$$

The bowl D^2 capping $M_{c_0-\varepsilon}$ from above is called a 2-*handle*. The relation (1.59) says that as t goes from $c_0 - \varepsilon$ to $c_0 + \varepsilon$, a 2-handle is attached to $M_{c_0-\varepsilon}$ to obtain $M_{c_0+\varepsilon}$. Note that in this process of attaching a 2-handle, one of the boundary circles of $M_{c_0-\varepsilon}$ is capped off, and the number of connected components of the boundary is reduced by one.

(d) Handle decompositions

We go back to a Morse function $f : M \to \mathbb{R}$.

By Lemma 1.21 the number of critical points of f is finite. We label these critical points as

$$p_1, \ p_2, \ p_3, \ \ldots, \ p_n.$$

We will show in the next chapter that we may perturb a given Morse function f in such a way that

$$f(p_i) \neq f(p_j), \quad \text{if} \quad i \neq j.$$

So we assume here that f satisfies this property. We change the subscripts of the p_i's if necessary and assume that

(1.60) $$f(p_1) < f(p_2) < f(p_3) < \cdots < f(p_n).$$

Setting $c_i = f(p_i)$, we get a sequence of numbers

(1.61) $$c_1 < c_2 < c_3 < \cdots < c_n.$$

It is evident that c_1 is the minimum value and c_n is the maximum value of f.

We trace how M_t changes as the parameter t increases, starting from somewhere less than c_1. As before, for $t < c_1$, we have $M_t = \emptyset$. As soon as t passes c_1, a disk (upright bowl) pops out and we have $M_t = D^2$. This is because the index of p_0 is zero, as c_1 is the minimum value of f. This 2-disk (upright bowl) corresponding to the critical point of index 0 is called a 0-*handle*. A 2-handle is also a 2-disk, but a 2-handle corresponds to an upside-down bowl and a 0-handle corresponds to an upright bowl. (Another interpretation of the index of a critical point: the number of downward directions or the dimension of the downward slope from the critical point is its index. As a 2-handle at a critical point is an upside-down bowl, there are two directions to go down from it and so its index is 2. As for a 1-handle, the center line of the bridge slopes downward from the critical point in question, which is one-dimensional. A 0-handle is an upright bowl, so there is no downward slope from this critical point.)

We continue the above process, and each time t passes a critical value c_i, we attach a 0-, 1-, or 2-handle to $M_{c_i-\varepsilon}$ depending on the index of the corresponding critical point p_i ("attaching" a 0-handle means taking the disjoint union of the handle and $M_{c_i-\varepsilon}$).

The last c_n is the maximum value of f, so it corresponds to a 2-handle. The boundary of $M_{c_n-\varepsilon}$ is a circle along which a 2-handle is attached to cap off $M_{c_n-\varepsilon}$ to complete the closed surface M. We have therefore shown the following theorem.

THEOREM 1.24 (Handle decomposition of a closed surface). *If a closed surface M admits a Morse function $f : M \to \mathbb{R}$, then M can be described as a union of finitely many 0-, 1-, and 2-handles.*

In the next chapter we show the existence of Morse functions on any closed surface M, so that, actually, any closed surface admits a handle decomposition.

Summary

1.1 A function admits two kinds of critical points: degenerate and non-degenerate.

1.2 A function which has only non-degenerate critical points is called a Morse function.

1.3 If f is a Morse function defined on a closed surface M, then we can slice off parts of M from below with respect to the values of f.

1.4 Each time the value of f passes through a critical value, a handle appears and is attached to the previously built-up subsurface. The index of the handle coincides with the index of the corresponding critical point.

1.5 In this way, a closed surface M can be decomposed into a union of finitely many handles.

Exercises

1.1 We obtain a closed surface from two disks D_1 and D_2 by pasting them together along their boundaries by a diffeomorphism $h : \partial D_1 \to \partial D_2$. Show that this surface is diffeomorphic to the 2-sphere S^2, using Lemma 1.20.

1.2 Think of the circle S^1 as the boundary of the disk D^2, and show that any homeomorphism $h : S^1 \to S^1$ can be extended to a homeomorphism $H : D^2 \to D^2$.

1.3 Under the same assumption as in the preceding question, show that any diffeomorphism $h : S^1 \to S^1$ can be extended to a diffeomorphism $H : D^2 \to D^2$.

1.4 We use angles θ to indicate points on the circle S^1, where θ and $\theta + 2\pi$ correspond to the same point on S^1. Define a function $f : S^1 \times S^1 \to \mathbb{R}$ on the torus $S^1 \times S^1$ by

$$f(\theta, \phi) = (R + r \cos \phi) \cos \theta,$$

where R and r are positive constants with $R > r$. Show that f is a Morse function, and find all of its critical points and their indices.

CHAPTER 2

Extension to General Dimensions

In this chapter we begin Morse theory for general m-dimensional manifolds. We extend the basic concepts that we explored in Chapter 1, such as non-degenerate critical points, Morse functions, handle decompositions and so on, to higher dimensions.

In Section 1, we give a quick review of manifolds, and in Section 2, we prove the existence of Morse functions on general manifolds. In Section 3 we discuss "gradient-like" vector fields of Morse functions. We prove that a Morse function can be modified slightly so that it takes distinct critical values at distinct critical points.

We will deal with general handle decompositions in Chapter 3.

2.1. Manifolds of dimension m

An m-dimensional manifold (m-manifold for short) is a generalization of the concept of surfaces to a higher dimension m. We refer the reader to textbooks on manifold theory (see [11], [17], [4] for example) for a detailed definition of a manifold. The only property of a manifold of class C^∞ we need here is the existence of local coordinate systems; that is, if M is an m-dimensional smooth (C^∞) manifold, then about each point p in M there exists a smooth (C^∞) coordinate system

$$(2.1) \qquad (x_1,\ x_2,\ \ldots,\ x_m).$$

In what follows we assume that manifolds and coordinate systems are smooth (C^∞).

(a) Functions on a manifold and maps between manifolds

We say that a function $f : M \to \mathbb{R}$ on M is *smooth*, or *of class* C^∞ (or simply C^∞), if at each point p in M, f is of class C^∞ with respect to a local coordinate system $(x_1, x_2, \ldots, x_m)$ about p. This definition is identical to the definition of a C^∞-function on a surface.

Let N be an n-manifold. We define when a continuous map $h : M \to N$ is smooth or of class C^∞. To do this, we first need to define when "$h : M \to N$ is of class C^∞ at a point $p \in M$." Choose small enough neighborhoods U and V of p and $h(p)$, respectively, such that

$$h(U) \subset V,$$

making sure that U and V are in some coordinate neighborhoods $(x_1, x_2, \ldots, x_m)$ and $(y_1, y_2, \ldots, y_n)$. Then we can locally write

$$(2.2) \qquad h(x_1, x_2, \ldots, x_m) = (y_1, y_2, \ldots, y_n)$$

about p. We say that "h is of class C^∞ about the point p" if h in equation (2.2) is of class C^∞ with respect to the local coordinate systems $(x_1, x_2, \ldots, x_m)$ and $(y_1, y_2, \ldots, y_n)$.

We look at equation (2.2) more carefully: each y_i on the right-hand side of this equation depends on $(x_1, x_2, \ldots, x_m)$ on the left-hand side, and hence we can think of it as a function of m variables, $x_1, x_2, \ldots, x_m$:

$$(2.3) \qquad y_i = h_i(x_1, x_2, \ldots, x_m).$$

The map h in equation (2.2) is C^∞ if and only if each of the n functions $h_i(x_1, x_2, \ldots, x_m)$ is C^∞. In this way we express the map h locally by n functions $h_1, h_2, \ldots, h_n$ of m variables. We say that $h_1, h_2, \ldots, h_n$ are *local components* of h.

Finally we define a *map $h : M \to N$ of class C^∞* to be a map $h : M \to N$ which is of class C^∞ at every point $p \in M$.

We often use the term "smooth" for class C^∞, just as we do with surfaces. The definitions of homeomorphisms and diffeomorphisms between manifolds are the same as those between surfaces.

(b) Manifolds with boundary

The idea of "manifolds with boundary" is just an extension of "surfaces with boundary" to more general dimensions. Let us look at some examples.

EXAMPLE 2.1 (The m-disk). We denote by $\mathbb{R}^m$ the Euclidean space of dimension m: $\mathbb{R}^m = \{\, (x_1, x_2, \ldots, x_m) \mid x_i \text{ is a real number} \,\}$. The *$m$-dimensional (unit) disk (m-disk) D^m* is the set of points in $\mathbb{R}^m$ which satisfy $x_1^2 + x_2^2 + \cdots + x_m^2 \leq 1$;

$$(2.4) \qquad D^m = \{\, (x_1, x_2, \ldots, x_m) \mid x_1^2 + x_2^2 + \cdots + x_m^2 \leq 1 \,\}.$$

The disk D^m is an m-dimensional manifold with boundary. The boundary of D^m is the $(m-1)$-*dimensional unit sphere* $((m-1)$-sphere)

$$(2.5) \qquad S^{m-1} = \{\, (x_1, x_2, \ldots, x_m) \mid x_1^2 + x_2^2 + \cdots + x_m^2 = 1 \,\}.$$

We use the boundary symbol ∂ and write this situation as

$$(2.6) \qquad\qquad \partial D^m = S^{m-1}$$

(see Figure 2.1).

In particular, the 3-dimensional disk D^3 is really a solid ball and its boundary ∂D^3 is the 2-dimensional sphere S^2, which is indeed a sphere. The m-dimensional disk is compact, but the next example is not.

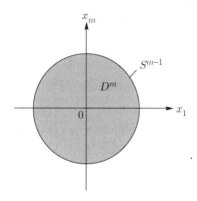

FIGURE 2.1. The m-disk and the $(m-1)$-sphere

EXAMPLE 2.2 (The m-dimensional upper half-space). Set

$$(2.7) \qquad\qquad \mathbb{R}_+^m = \{\, (x_1, x_2, \ldots, x_m) \mid x_m \geq 0 \,\}.$$

We say that $\mathbb{R}_+^m$ is the m-*dimensional upper half-space*. The boundary of $\mathbb{R}_+^m$ is defined by $x_m = 0$, and we identify it with $\mathbb{R}^{m-1} = \{\, (x_1, x_2, \ldots, x_{m-1}) \,\}$ (see Figure 2.2);

$$(2.8) \qquad\qquad \partial \mathbb{R}_+^m = \mathbb{R}^{m-1}.$$

We now generalize the above two examples as follows.

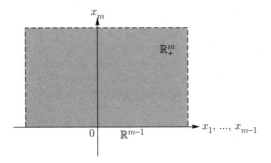

FIGURE 2.2. The m-dimensional upper half-space

Let M be an m-manifold (without boundary) and let $f : M \to \mathbb{R}$ be a smooth function on M. Assume that 0 is not a critical value of f (see Definition 2.10). Define a subset $M_{f \geq 0}$ of M by

$$(2.9) \qquad M_{f \geq 0} = \{\, p \in M \mid f(p) \geq 0 \,\};$$

then $M_{f \geq 0}$ is an m-manifold with boundary, and its boundary $\partial M_{f \geq 0}$ is the set

$$(2.10) \qquad M_{f = 0} = \{\, p \in M \mid f(p) = 0 \,\}.$$

In fact, we can define the disk D^m or the upper-half space $\mathbb{R}^m_+$ by taking $\mathbb{R}^m$ as M and defining $f : \mathbb{R}^m \to \mathbb{R}$ by

$$(2.11) \qquad f = \begin{cases} 1 - (x_1^2 + x_2^2 + \cdots + x_m^2) & \text{for } D^m, \\ x_m & \text{for } \mathbb{R}^m_+. \end{cases}$$

Throughout the rest of this book we only use manifolds with boundary which we can express as $M_{f \geq 0}$ (in fact, if we define "manifolds with boundary" rigorously, we can show that they all have the form of $M_{f \geq 0}$).

The next theorem guarantees that the boundary $M_{f=0}$ is an $(m-1)$-manifold.

THEOREM 2.3 (The implicit function theorem). *Let M be an m-manifold (in the usual sense) and let $f : M \to \mathbb{R}$ be a smooth function defined on M. If 0 is not a critical value of f, then the subset $f^{-1}(0) = \{p \in M \mid f(p) = 0\}$ of M is an $(m-1)$-dimensional submanifold of M.*

The definition of a "submanifold" is as follows.

DEFINITION 2.4 (Submanifolds). Let M be a manifold of dimension m. We say that a subset K of M is a k-*dimensional submanifold* of M if for every point p of K, there exists a (C^∞) local coordinate system about p of M

$$(x_1, x_2, \ldots, x_m)$$

such that K is described by the equations

(2.12) $$x_{k+1} = x_{k+2} = \cdots = x_m = 0$$

within this coordinate neighborhood (see Figure 2.3).

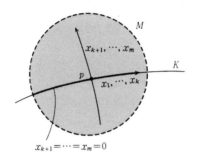

FIGURE 2.3. A k-dimensional submanifold

It follows from Definition 2.4 of submanifolds that every point of K admits a local coordinate system, and hence K itself is a k-dimensional manifold.

The major point in the proof of the implicit function theorem is to show that if 0 is not a critical value of f, then we can construct about each point p of $f^{-1}(0)$ a local coordinate system in M having f as the m-th coordinate:

(2.13) $$(x_1, x_2, \ldots, x_m), \quad \text{where } x_m = f.$$

Once we show this, we can express $f^{-1}(0)$ by the equation x_m ($= f$) $= 0$ near each point p in a suitable coordinate system, and so $f^{-1}(0)$ is an $(m-1)$-submanifold of M according to the definition (of submanifolds). We refer the reader to the textbook [11] (or any calculus book) for a detailed proof.

With the local coordinate system $(x_1, x_2, \ldots, x_m)$ of (2.13) about p, the subsets $M_{f\geq 0}$ and $M_{f=0}$ have local expressions $x_m \geq 0$ and $x_m = 0$, respectively. We thus have the following lemma, where $M_{f\geq 0}$ is replaced by M.

LEMMA 2.5. (Local coordinate systems about boundary points).
*Let M be an m-manifold with boundary, and let p be an arbitrary point
of the boundary ∂M. Then there exists a local coordinate system*

$$(2.14) \qquad (x_1, x_2, \ldots, x_m), \quad x_m \geq 0,$$

*about p. Notice that M corresponds to $x_m \geq 0$ and ∂M corresponds
to $x_m = 0$ (see Figure 2.4).*

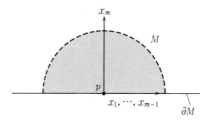

FIGURE 2.4. A local coordinate system of the upper
half-plane type

In what follows, we assume that a local coordinate system about
a point p of the boundary is always of this type of upper half-plane.
We actually have a stronger result than the above lemma: the
boundary ∂M of a manifold with boundary M has a *collar neighborhood* of the form $\partial M \times [0, 1)$, where $[0, 1)$ is the right half-open interval
$t \in \{\mathbb{R} \mid 0 \leq t < 1\}$. The boundary ∂M corresponds to $\partial M \times \{0\}$.
We will discuss this in Section 3 (cf. Corollary 2.33, Section 3). The
term "collar" stands for the collar of a shirt.

(c) Functions and maps on manifolds with boundary

We used the term C^∞ for functions and maps on surfaces with
boundary rather carelessly and intuitively in Chapter 1. We now rede-
fine smooth functions and smooth maps on manifolds with boundary.
For we must treat these maps with care around the boundary points.
In Definition 2.6, M is a manifold with boundary and $f : M \to \mathbb{R}$ is
a function on M. As in the case of surfaces, we denote by $\mathrm{int}(M)$ the
complememt of the boundary in M:

$$(2.15) \qquad \mathrm{int}(M) = M - \partial M.$$

We call $\mathrm{int}(M)$ the *interior* of M.

DEFINITION 2.6. A function $f : M \to \mathbb{R}$ is *smooth* (of class C^∞) at a point p ($\in M$) if one of the following conditions (i) and (ii) holds.

(i) If p is an interior point, f is smooth with respect to a local coordinate system $(x_1, x_2, \ldots, x_m)$ in a suitably small neighborhood of p.

(ii) If p is a boundary point and we express f with respect to a local coordinate system $(x_1, x_2, \ldots, x_m)$ with $x_m \geq 0$ in a small enough neighborhood of p, then $f(x_1, x_2, \ldots, x_m)$ can be extended to a smooth function of m variables

$$\tilde{f}(x_1, x_2, \ldots, x_m)$$

defined with respect to the coordinate system $(x_1, x_2, \ldots, x_m)$, $x_i \in \mathbb{R}$ for all i (that is, $\tilde{f}$ is defined without the restriction $x_m \geq 0$). In other words, f is the restriction of $\tilde{f}$ to $x_m \geq 0$. A more intuitive expression might be $f = \tilde{f} \mid \{x_m \geq 0\}$.

We say that $f : M \to \mathbb{R}$ is *smooth* if it is smooth at every point p of M.

We now define a smooth map $h : M \to N$ between manifolds M and N with boundary. First, as in the case without boundary, we say that "h is smooth at a point p" if each component of the coordinate representation $h_1, h_2, \ldots, h_n$ of h in a neighborhood of p is smooth, in the sense defined in Definition 2.6. If $h : M \to N$ is smooth at every point of M, then we say that h is *smooth*. This completes the definition of smooth maps between manifolds with boundary.

A homeomorphism $h : M \to N$ is a *diffeomorphism* if both $h : M \to N$ and $h^{-1} : N \to M$ are smooth in the way we have just defined (this is also analogous to the case of manifolds without boundary).

A diffeomorphism $h : M \to N$ maps the boundary ∂M of M onto the boundary ∂N of N. The restriction of h

$$h \mid \partial M : \partial M \to \partial N$$

is a diffeomorphism from ∂M to ∂N.

We used the following theorem in Chapter 1.

THEOREM 2.7 (Gluing manifolds with boundary). *Let M_1 and M_2 be manifolds with boundary, and let $\varphi : \partial M_1 \to \partial M_2$ be a diffeomorphism between the boundaries. Then we can construct a new manifold $W = M_1 \cup_\varphi M_2$ by gluing the boundaries of M and N using φ (that is, by identifying each point $p \in \partial M_1$ with the point $\varphi(p) \in M_2$).*

The resulting manifold W is unique up to diffeomorphism. (It is allowed to glue only a part of connected components of the boundary, instead of the entire boundary. See Figure 2.5).

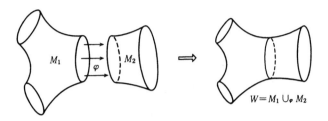

FIGURE 2.5. Gluing manifolds with boundary

We refer the reader to [14] for a proof.

Next we discuss gluing diffeomorphisms, which is slightly more delicate than gluing manifolds.

THEOREM 2.8 (Gluing diffeomorphisms). *Let $W = M_1 \cup_\varphi M_2$ and $V = N_1 \cup_\psi N_2$ be the manifolds obtained by gluing manifolds with boundary (where $\varphi : \partial M_1 \to \partial M_2$ and $\psi : \partial N_1 \to \partial N_2$ are diffeomorphisms). Suppose that we have diffeomorphisms $h_1 : M_1 \to N_1$ and $h_2 : M_2 \to N_2$ such that $\psi \circ h_1(p) = h_2 \circ \varphi(p)$ for every point p in ∂M_1. Then there exists a diffeomorphism $H = h_1 \cup h_2 : W \to V$ obtained by gluing h_1 and h_2 along the boundary (see Figure 2.6).*

The notation $H = h_1 \cup h_2$ in this theorem is a bit deceiving. The map H may not quite restrict to h_1 on M_1 nor h_2 on M_2, since if we take such a map as H, then the resulting map from W to V may not be differentiable along ∂M_1, and so we may have to modify h_1 and h_2 near the respective boundaries. We do not have enough space to explain this modification in detail in this book, but we hope that the following example will convince the reader.

EXAMPLE 2.9. Let $M_1 = N_1 = \mathbb{R}_+^2 = \{(x,y) \mid y \geq 0\}$ be the upper half-plane and let $M_2 = N_2 = \mathbb{R}_-^2 = \{(x,y) \mid y \leq 0\}$ be the lower half-plane. Let φ and ψ be the identity maps so that $W = V = \mathbb{R}^2$. Define maps h_1 and h_2 by

$$(2.16) \qquad \begin{cases} h_1(x,y) = (x+y, y) & \text{(if } y \geq 0), \\ h_2(x,y) = (x,y) & \text{(if } y \leq 0). \end{cases}$$

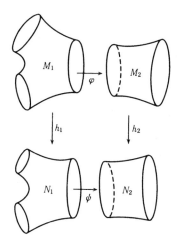

FIGURE 2.6. Gluing diffeomorphisms

Then both $h_1 : M_1 \to N_1$ and $h_2 : M_2 \to N_2$ are diffeomorphims; however, simply putting these maps together does not yield a diffeomorphism of $\mathbb{R}^2$ onto $\mathbb{R}^2$. Thus we modify h_1 as follows:

$$(2.17) \qquad \tilde{h}_1(x,y) = (x + \rho(y)y, \; y),$$

where (for a small enough $\varepsilon > 0$) $\rho(y)$ is a smooth function such that $0 \leq \rho(y) \leq 1$, $\rho(y) = 0$ for $y \leq \varepsilon$, and $\rho(y) = 1$ for $y \geq 2\varepsilon$ (see [17] for the existence of such a function ρ). With this modification, the function $H : \mathbb{R}^2 \to \mathbb{R}^2$ defined to be $\tilde{h}_1$ on the upper half-plane and h_2 on the lower half-plane will be a diffeomorphism. In Theorem 2.8, we simply wrote $H = h_1 \cup h_2$ with this kind of H in mind.

2.2. Morse functions

(a) Morse functions on m-manifolds

Let M be an m-manifold without boundary and $f : M \to \mathbb{R}$ a smooth function defined on it.

DEFINITION 2.10 (Critical points of f). A point p_0 of M is a *critical point* of f if we have

$$(2.18) \qquad \frac{\partial f}{\partial x_1}(p_0) = 0, \; \frac{\partial f}{\partial x_2}(p_0) = 0, \; \cdots, \; \frac{\partial f}{\partial x_m}(p_0) = 0,$$

with respect to a local coordinate system $(x_1, x_2, \ldots, x_m)$ about p_0.

This definition does not depend on the choice of a coordinate system; that is, if the condition (2.18) holds for $(x_1, x_2, \ldots, x_m)$, then it holds for any other coordinate system $(y_1, y_2, \ldots, y_m)$ (Exercise 2.1).

A real number c is a *critical value* of $f : M \to \mathbb{R}$ if $f(p_0) = c$ for some critical point p_0 of f. We have used this term in the preceding section already in the definition of manifolds with boundary.

DEFINITION 2.11 (Hessian). Let p_0 be a critical point of $f : M \to \mathbb{R}$. We define the *Hessian* of the function f at the critical point p_0 to be the $m \times m$ matrix

(2.19)

$$
H_f(p_0) = \begin{pmatrix}
\frac{\partial^2 f}{\partial x_1^2}(p_0) & & \cdots & & \frac{\partial^2 f}{\partial x_1 \partial x_m}(p_0) \\
& \ddots & & & \\
\vdots & & \frac{\partial^2 f}{\partial x_i \partial x_j}(p_0) & & \vdots \\
& & & \ddots & \\
\frac{\partial^2 f}{\partial x_m \partial x_1}(p_0) & & \cdots & & \frac{\partial^2 f}{\partial x_m^2}(p_0)
\end{pmatrix}.
$$

The (i, j)-entry of the Hessian of f is $\frac{\partial^2 f}{\partial x_i \partial x_j}(p_0)$, and since

$$
\frac{\partial^2 f}{\partial x_i \partial x_j}(p_0) = \frac{\partial^2 f}{\partial x_j \partial x_i}(p_0),
$$

the Hessian of f is a symmetric matrix.

We compute the second order partial derivatives of f with respect to a new coordinate system $(y_1, y_2, \ldots, y_m)$ and compare them with those with respect to the original coordinate system $(x_1, x_2, \ldots, x_m)$, and obtain

(2.20) $$
\frac{\partial^2 f}{\partial y_h \partial y_k}(p_0) = \sum_{i,j=1}^{m} \frac{\partial x_i}{\partial y_h}(p_0) \frac{\partial x_j}{\partial y_k}(p_0) \frac{\partial^2 f}{\partial x_i \partial x_j}(p_0).
$$

We thus have the following lemma.

LEMMA 2.12. *Consider two coordinate systems*

$$(y_1, y_2, \ldots, y_m) \quad and \quad (x_1, x_2, \ldots, x_m)$$

at a critical point p_0, and let $\mathcal{H}_f(p_0)$ and $H_f(p_0)$ be the Hessians of f with respect to these coordinate systems, respectively. Then $\mathcal{H}_f(p_0)$ and $H_f(p_0)$ are related as

$$(2.21) \qquad \mathcal{H}_f(p_0) = {}^t\! J(p_0) H_f(p_0) J(p_0),$$

where $J(p_0)$ is the Jacobian (matrix) of the coordinate transformation from $(y_1, y_2, \ldots, y_m)$ to $(x_1, x_2, \ldots, x_m)$ evaluated at p_0:

$$(2.22) \qquad J(p_0) = \begin{pmatrix} \frac{\partial x_1}{\partial y_1}(p_0) & & \cdots & & \frac{\partial x_1}{\partial y_m}(p_0) \\ & \ddots & & & \\ \vdots & & \frac{\partial x_i}{\partial y_j}(p_0) & & \vdots \\ & & & \ddots & \\ \frac{\partial x_m}{\partial y_1}(p_0) & & \cdots & & \frac{\partial x_m}{\partial y_m}(p_0) \end{pmatrix}.$$

We now define non-degenerate and degenerate critical points.

DEFINITION 2.13 (Non-degenerate & degenerate critical points). We say that a critical point p_0 is *non-degenerate* if the determinant $\det H_f(p_0)$ of the Hessian of f at p_0 is not zero, and we say that it is *degenerate* if $\det H_f(p_0) = 0$.

As a corollary to Lemma 2.12, we have

COROLLARY 2.14. *The property of a critical point p_0 of a function $f : M \to \mathbb{R}$ being non-degenerate or degenerate does not depend on the choice of a coordinate system at p_0.*

PROOF. We have

$$\det \mathcal{H}_f(p_0) = \det {}^t\! J(p_0) \det H_f(p_0) \det J(p_0)$$

from the relation (2.21). Since the Jacobian $J(p_0)$ of the coordinate transformation at p_0 has a non-zero determinant, the above relation implies that $\det \mathcal{H}_f(p_0) \neq 0$ if and only if $\det H_f(p_0) \neq 0$. $\qquad\square$

We see that all the above definitions for manifolds of general dimensions are generalizations of those we discussed for the surface case. One last instance is the definition of a Morse function.

DEFINITION 2.15 (Morse function). We say that a function $f : M \to \mathbb{R}$ is a *Morse function* if every critical point of f is non-degenerate.

(b) The Morse lemma for dimension m

We extend the Morse lemma (Theorem 1.11) for surfaces to the case of manifolds.

THEOREM 2.16 (Morse lemma for dimension m). *Let p_0 be a non-degenerate critical point of $f : M \to \mathbb{R}$. Then we can choose a local coordinate system $(X_1, X_2, \ldots, X_m)$ about p_0 such that the co-ordinate representation of f with respect to these coordinates has the following standard form:*

$$(2.23) \qquad f = -X_1^2 - X_2^2 - \cdots - X_\lambda^2 + X_{\lambda+1}^2 + \cdots + X_m^2 + c,$$

where p_0 corresponds to the origin $(0, 0, \ldots, 0)$ and c is a constant $(= f(p_0))$.

The number λ of minus signs in the standard form is the number of negative diagonal entries of the Hessian $H_f(p_0)$ after diagonalization. According to Sylvester's law (see [20]), λ does not depend on the way the Hessian is diagonalized, so that λ is determined by the function f and the critical point p_0.

DEFINITION 2.17. The number λ is called the *index* of a non-degenerate critical point p_0. The index of p_0 is an integer between (including) 0 and m (the dimension of M).

We now prove the Morse lemma (Theorem 2.16) for dimension m.

PROOF. Choose a local coordinate system $(x_1, x_2, \ldots, x_m)$ at the critical point p_0, where p_0 corresponds to the origin $(0, 0, \ldots, 0)$. We may further assume that $f(p_0) = 0$, replacing f by $f - f(p_0)$ if necessary.

Taking for granted the m-dimensional version of the "fundamental fact in calculus", which we used in the proof of Morse lemma for dimension two (Theorem 1.11), under the assumption that $f(0, 0, \ldots, 0) = 0$, we obtain m smooth functions (defined in a neighborhood of the origin)

$$(2.24) \qquad g_1(x_1, \ldots, x_m), \ g_2(x_1, \ldots, x_m), \ldots, g_m(x_1, \ldots, x_m)$$

such that

$$(2.25) \qquad f(x_1, \ldots, x_m) = \sum_{i=1}^{m} x_i g_i(x_1, \ldots, x_m)$$

and

(2.26)
$$\frac{\partial f}{\partial x_i}(0,\ldots,0) = g_i(0,\ldots,0).$$

Since the point $p_0 = (0,\ldots,0)$ is a critical point, both sides of (2.26) turn out to be zero. Hence, we can apply the "fundamental fact in calculus" to $g_i(x_1,\ldots,x_m)$ and find m smooth functions (defined in a neighborhood of the origin)

(2.27) $h_{i1}(x_1,\ldots,x_m),\ h_{i2}(x_1,\ldots,x_m),\ldots,\ h_{im}(x_1,\ldots,x_m)$

such that

(2.28)
$$g_i(x_1,\ldots,x_m) = \sum_{j=1}^{m} x_j h_{ij}(x_1,\ldots,x_m)$$

in a neighborhood of the origin. We plug this equation into (2.25) to get

(2.29)
$$f(x_1,\ldots,x_m) = \sum_{i,j=1}^{m} x_i x_j h_{ij}(x_1,\ldots,x_m).$$

Setting $H_{ij} = (h_{ij} + h_{ji})/2$, we get

(2.30)
$$f(x_1,\ldots,x_m) = \sum_{i,j=1}^{m} x_i x_j H_{ij}(x_1,\ldots,x_m)$$

and

(2.31)
$$H_{ij}(x_1,\ldots,x_m) = H_{ji}(x_1,\ldots,x_m).$$

We shall refer to the representation of $f(x_1,\ldots,x_m)$ in (2.30) as a "quadratic form representation," or a "representation by a quadratic form," of f. Note that the standard form (2.23) of f which we are to show is also a special type of a quadratic representation of f.

The idea of the proof is to change the representation of f in (2.30) to a quadratic representation in a standard form of Morse type, by induction on the number of terms involved in the quadratic form representing f.

Let us now compute second order partial derivatives of (2.30) at the origin. We obtain

(2.32)
$$\frac{\partial^2 f}{\partial x_i \partial x_j}(0,\ldots,0) = 2H_{ij}(0,\ldots,0).$$

Just as we saw in the proof of the 2-dimensional Morse lemma (1.11), if we use the assumption that the critical point p_0 is non-degenerate ($\det H_{ij}(0,\ldots,0) \neq 0$), we may assume that

$$(2.33) \qquad \frac{\partial^2 f}{\partial x_1^2}(0,\ldots,0) \neq 0,$$

after a suitable linear transfomation of the local coordinate system $(x_1,\ldots,x_m)$. Then by (2.32), $H_{11}(0,\ldots,0) \neq 0$. Since the function H_{11} is continuous, we see that

$$(2.34) \qquad H_{11} \text{ is not 0 in a neighborhood of the origin.}$$

Now we introduce a new coordinate system $(X_1, x_2, \ldots, x_m)$, where we define X_1 by

$$(2.35) \qquad X_1 = \sqrt{|H_{11}|}\left(x_1 + \sum_{i=2}^{m} x_i \frac{H_{1i}}{H_{11}}\right).$$

We can easily compute and show that the determinant of the Jacobian of the transformation from $(X_1, x_2, \ldots, x_m)$ to $(x_1, x_2, \ldots, x_m)$ at the origin is not zero, so that $(X_1, x_2, \ldots, x_m)$ is certainly a local coordinate system. The square of X_1 is computed as follows:

$$(2.36) \qquad X_1^2 = |H_{11}|\left(x_1 + \sum_{i=2}^{m} x_i \frac{H_{1i}}{H_{11}}\right)^2$$

$$= \begin{cases} H_{11}x_1^2 + 2\sum_{i=2}^{m} x_1 x_i H_{1i} + \left(\sum_{i=2}^{m} x_i H_{1i}\right)^2/H_{11} & (H_{11} > 0), \\ -H_{11}x_1^2 - 2\sum_{i=2}^{m} x_1 x_i H_{1i} - \left(\sum_{i=2}^{m} x_i H_{1i}\right)^2/H_{11} & (H_{11} < 0). \end{cases}$$

Comparing (2.36) with the quadratic form representation (2.30) of f, we see that

$$(2.37)$$

$$f = \begin{cases} X_1^2 + \sum_{i,j=2}^{m} x_i x_j H_{ij} - \left(\sum_{i=2}^{m} x_i H_{1i}\right)^2/H_{11} & (H_{11} > 0), \\ -X_1^2 + \sum_{i,j=2}^{m} x_i x_j H_{ij} - \left(\sum_{i=2}^{m} x_i H_{1i}\right)^2/H_{11} & (H_{11} < 0). \end{cases}$$

In (2.37), the second term, and after, are a sum over $x_2, x_3, \ldots, x_m$, so that this part is simplified to a quadratic form representation of fewer variables than the representation (2.30). Now we may proceed by induction on the number of variables to prove that we can represent f as a standard Morse form.   □

From the Morse lemma, one obtains the following results, just as in the surface case.

COROLLARY 2.18. *A non-degenerate critical point is isolated.*

COROLLARY 2.19. *A Morse function defined on a compact manifold admits only finitely many critical points.*

(c) Existence of Morse functions
Until now we have pursued analogues of the surface case for higher-dimensional manifolds. Now we show something new; we prove the existence theorem for Morse functions. By a *closed manifold* we mean a compact manifold without boundary.

THEOREM 2.20 (Existence of Morse functions). *Let M be a closed m-manifold and let $g : M \to \mathbb{R}$ be a smooth function defined on M. Then there exists a Morse function $f : M \to \mathbb{R}$ arbitrarily close to $g : M \to \mathbb{R}$.*

Since there are many smooth functions on M, this theorem implies that there are many Morse functions defined on M as well. (For example, a constant function which assigns a constant number c_0 to every point of M is certainly smooth, so that there is a Morse function $f : M \to \mathbb{R}$ close to it! Of course f is not a constant function.)
In the theorem "arbitrarily close to g" means "arbitrarily C^2-close." The meaning of this will become clear in the proof.
We need the following two lemmas for the proof of the existence theorem.

LEMMA 2.21. *Let $\mathbb{R}^m = \{(x_1, x_2, \ldots, x_m)\}$ be m-dimensional Euclidean space, U an open set of $\mathbb{R}^m$, and $f : U \to \mathbb{R}$ a smooth function on U. Then for some real numbers $a_1, a_2, \ldots, a_m$,*

$$(2.38) \qquad f(x_1, x_2, \ldots, x_m) - (a_1 x_1 + a_2 x_2 + \cdots + a_m x_m)$$

is a Morse function on U. Moreover, one can choose $a_1, a_2, \ldots, a_m$ in such a way that their absolute values are as small as one wishes.

First let us describe an "intuitive proof" of this lemma. We only need to show that the critical points of the function (2.38) are non-degenerate. We ask where a critical point $p_0 = (x_1^0, x_2^0, \ldots, x_m^0)$ of the function (2.38) can possibly exist: that happens at any point p_0 where the "tangent plane" to the graph of the function

$$y = f(x_1, x_2, \ldots, x_m)$$

at the point p_0 is parallel to the graph of the linear function

$$y = a_1 x_1 + a_2 x_2 + \cdots + a_m x_m.$$

When can a point p_0 be a degenerate critical point? That happens when the tangent plane to the graph of the function $y = f(x_1, x_2, \ldots, x_m)$ meets the graph with a high degree of contact (here "high" means higher than two). On the other hand, there would not be many tangent planes meeting the graph with such high degrees of contact, so that for almost any choice of numbers $a_1, a_2, \ldots, a_m$, the graph of the function $y = a_1 x_1 + \cdots + a_m x_m$ would not be parallel to such tangent planes. Then each critical point of the above difference function is non-degenerate. This is the intuitive proof.

We need Sard's theorem to prove the lemma rigorously.

First we give some preliminary explanations. Let U be an open set in $\mathbb{R}^m$ and consider a smooth map

$$(2.39) \qquad h : U \to \mathbb{R}^m.$$

The map h sends a point $(x_1, x_2, \ldots, x_m)$ of U to a point $(y_1, y_2, \ldots, y_m)$ of $\mathbb{R}^m$, where each component y_i has the expression

$$y_i = h_i(x_1, x_2, \ldots, x_m), \quad i = 1, 2, \cdots, m,$$

as a function of $(x_1, x_2, \ldots, x_m)$. Recall that this is the coordinate representation of the map $h : U \to \mathbb{R}^m$. For convenience we write the coordinate representation of $h : U \to \mathbb{R}^m$ as a column vector

$$(2.40) \qquad h = \begin{pmatrix} h_1 \\ h_2 \\ \vdots \\ h_m \end{pmatrix}.$$

We then define the *Jacobian (matrix)* of h at a point $p_0 \ (\in U)$ to be the following $m \times m$ matrix:

$$(2.41) \qquad J_h(p_0) = \begin{pmatrix} \frac{\partial h_1}{\partial x_1}(p_0) & & \cdots & & \frac{\partial h_1}{\partial x_m}(p_0) \\ & \ddots & & & \\ \vdots & & \frac{\partial h_i}{\partial x_j}(p_0) & & \vdots \\ & & & \ddots & \\ \frac{\partial h_m}{\partial x_1}(p_0) & & \cdots & & \frac{\partial h_m}{\partial x_m}(p_0) \end{pmatrix}.$$

DEFINITION 2.22. (Critical points and critical values of a map). A point p_0 of U with $\det J_h(p_0) = 0$ is called a *critical point* of a map $h : U \to \mathbb{R}^m$. The image $h(p_0)$ in $\mathbb{R}^m$ of a critical point p_0 of $h : U \to \mathbb{R}^m$ is a *critical value* of h.

Note that this definition works only when the dimension of U is m (which is the dimension of $\mathbb{R}^m$). For maps between manifolds of distinct dimensions, we need to define critical points in a different way. Compare [11].

We state Sard's theorem as follows.

THEOREM 2.23 (Sard's theorem). *The set of critical values of a smooth map $h : U \to \mathbb{R}^m$ has measure zero in $\mathbb{R}^m$.*

We omit a precise definition of "measure zero" here, and only mention that having "measure zero" for $m = 2$ is basically the same as having zero area, and having "measure zero" for $m = 3$ is the same as having zero volume. In short, Sard's theorem says that "there are not many" critical values in $\mathbb{R}^m$. In particular, there is always a point which is **not** a critical value of $h : U \to \mathbb{R}^m$ in any neighborhood of any point in $\mathbb{R}^m$. We refer the reader to [11], [15] for the proof of Sard's theorem. We now give a rigorous proof of Lemma 2.21.

PROOF. Our goal is to select small enough $a_1, a_2, \ldots, a_m$ to make

$$(2.42) \qquad f(x_1, x_2, \ldots, x_m) - (a_1 x_1 + a_2 x_2 + \cdots + a_m x_m)$$

a Morse function.

First we define a map $h : U \to \mathbb{R}^m$ by

$$(2.43) \qquad h = \begin{pmatrix} \frac{\partial f}{\partial x_1} \\ \frac{\partial f}{\partial x_2} \\ \vdots \\ \frac{\partial f}{\partial x_m} \end{pmatrix},$$

using the function $f : U \to \mathbb{R}$. This is the map whose i-th component is the i-th partial derivative of f. The Jacobian of $h : U \to \mathbb{R}^m$ at the point p_0 is

$$(2.44)$$

$$J_h(p_0) = \begin{pmatrix} \frac{\partial^2 f}{\partial x_1^2}(p_0) & \cdots & & & \frac{\partial^2 f}{\partial x_1 \partial x_m}(p_0) \\ & \ddots & & & \\ \vdots & & \frac{\partial^2 f}{\partial x_i \partial x_j}(p_0) & & \vdots \\ & & & \ddots & \\ \frac{\partial^2 f}{\partial x_m \partial x_1}(p_0) & \cdots & & & \frac{\partial^2 f}{\partial x_m^2}(p_0) \end{pmatrix},$$

which is equal to the Hessian $H_f(p_0)$. Hence, p_0 is a critical point of the map $h : U \to \mathbb{R}^m$ if and only if $\det H_f(p_0) = 0$.

Choose a point of $\mathbb{R}^m$,

$$(2.45) \qquad \begin{pmatrix} a_1 \\ a_2 \\ \vdots \\ a_m \end{pmatrix},$$

which is *not* a critical value of $h : U \to \mathbb{R}^m$. By Sard's theorem, there are abundant such points, and moreover we can choose $a_1, a_2, \ldots, a_m$ with arbitrarily small absolute values. These a_i's are what we want.

Now we show that

$$\tilde{f}(x_1, x_2, \ldots, x_m) = f(x_1, x_2, \ldots, x_m) - (a_1 x_1 + a_2 x_2 + \cdots + a_m x_m)$$

is a Morse function on U.

If p_0 is a critical point of the function $\tilde{f}$, then since

$$(2.46) \qquad \frac{\partial \tilde{f}}{\partial x_i}(p_0) = \frac{\partial f}{\partial x_i}(p_0) - a_i = 0, \quad i = 1, 2, \ldots, a_m,$$

we have

$$(2.47) \qquad h(p_0) = \begin{pmatrix} a_1 \\ a_2 \\ \vdots \\ a_m \end{pmatrix},$$

by the definition of the map $U \to \mathbb{R}^m$. We have, however, chosen the a_i's so that

$$\begin{pmatrix} a_1 \\ a_2 \\ \vdots \\ a_m \end{pmatrix}$$

is not a critical value of $h : U \to \mathbb{R}^m$. Thus p_0 is not a critical point of $h : U \to \mathbb{R}^m$, and hence, as we noted earlier, we have $\det H_f(p_0) \neq 0$. Since f and $\tilde{f}$ differ only by a linear function, their Hessians coincide:

$$(2.48) \qquad H_f(p_0) = H_{\tilde{f}}(p_0).$$

Therefore, $\det H_f(p_0) \neq 0$ if and only if $\det H_{\tilde{f}}(p_0) \neq 0$, so that p_0 is a non-degenerate critical point of $\tilde{f}$. This completes the proof of Lemma 2.21.   □

Before proving the existence theorem (Theorem 2.20), we clarify
what we mean by "two functions are close," or more precisely, when
two smooth functions $f, g : M \to \mathbb{R}$ are "C^2-close," which appeared
in the statement of the theorem.

We say that an open set U is a *coordinate neighborhood* when U
has a local coordinate system $(x_1, x_2, \ldots, x_m)$ defined on it. When
we talk about a coordinate neighborhood U, we assume that a local
coordinate system $(x_1, x_2, \ldots, x_m)$ defined on U is specified.

We first define when f and g are C^2-close on a compact set K
(for instance, an m-disk of a suitable radius) contained in a coordinate
neighborhood. Let $\varepsilon > 0$ be a positive number. We say that "f is
a (C^2, ε)-approximation of g in K" if the following three inequalities
hold at every point p in K:

$$(2.49) \quad \begin{cases} |f(p) - g(p)| < \varepsilon, \\ \left| \frac{\partial f}{\partial x_i}(p) - \frac{\partial g}{\partial x_i}(p) \right| < \varepsilon, & i = 1, 2, \ldots, m, \\ \left| \frac{\partial^2 f}{\partial x_i \partial x_j}(p) - \frac{\partial^2 g}{\partial x_i \partial x_j}(p) \right| < \varepsilon, & i, j = 1, 2, \ldots, m. \end{cases}$$

In order to define a (C^2, ε)-approximation on a manifold M, we
cover M with a finite number of coordinate neighborhoods. For a
general manifold one may not be able to find such a finite cover, but
it is always possible to do so if M is compact, due to the definition of
compactness. We now take the opportunity to give a formal definition
of compactness.

DEFINITION 2.24 (Compactness). A topological space X is *compact* if among any infinite number of open sets $U_\alpha, U_\beta, \ldots, U_\lambda, \ldots$
which cover X:

$$(2.50) \qquad X = U_\alpha \cup U_\beta \cup \cdots \cup U_\lambda \cup \cdots,$$

there exist a finite number of open sets $U_\alpha, U_\beta, \ldots, U_\gamma$ which still
cover X:

$$(2.51) \qquad X = U_\alpha \cup U_\beta \cup \cdots \cup U_\gamma.$$

In short, X is compact if one can select a finite cover out of an infinite
cover of X.

According to a theorem in calculus (the Heine–Borel theorem),
any bounded closed set in $\mathbb{R}^m$ is compact (a set X is bounded in $\mathbb{R}^m$ if
it is contained within a certain distance from the origin). For instance,
the m-disk D^m and the $(m-1)$-sphere S^{m-1} are both compact.

We can certainly cover any manifold M with an infinite number of coordinate neighborhoods. If we assume that M is compact, then we can select a finite number of coordinate neighborhoods U_1, U_2, ..., U_k among them so that

$$(2.52) \qquad M = U_1 \cup U_2 \cup \cdots \cup U_k.$$

Now back to the definition of C^2-closeness: we cover M with a finite number of coordinate neighborhoods U_1, U_2, ..., U_k. Furthermore, we choose for each $i = 1, 2, \ldots, k$ a compact set K_i in U_i such that M is the union of the K_i's:

$$(2.53) \qquad M = K_1 \cup K_2 \cup \cdots \cup K_k.$$

REMARK. We can choose such K_i's as follows. We first consider all possible coordinate neighborhoods U in M and all possible pairs (U, D) where D runs through all possible m-disks contained in U. It is evident that the interiors $\text{int}(D)$ (which are open) of these disks D cover M. As M is compact, one can cover it with a finite number of them: $\text{int}(D_1)$, $\text{int}(D_2)$, ..., $\text{int}(D_k)$. Then the coordinate neighborhoods U_i's paired with respective D_i cover M. Setting $D_l = K_l$, we see that the compact sets K_1, K_2, ..., K_k contained in U_1, U_2, ..., U_k also cover M.

From now on, whenever we discuss (C^2, ε)-approximation, we assume that a finite cover $M = U_1 \cup \cdots \cup U_k$ of M, and a finite compact cover $M = K_1 \cup \cdots \cup K_k$ with $K_i \subset U_i$, are chosen and fixed.

With the above preliminaries, we give the following definition.

DEFINITION 2.25. A function $f : M \to \mathbb{R}$ is a (C^2, ε)-*approximation* of a function $g : M \to \mathbb{R}$ if f is a (C^2, ε)-approximation of g on K_l for each $l = 1, 2, \ldots, k$.

We need one more lemma for the proof of Theorem 2.20.

LEMMA 2.26. *Let C be a compact set in an m-dimensional manifold M. Suppose that a function $g : M \to \mathbb{R}$ has no degenerate critical point in C. Then for a sufficiently small number $\varepsilon > 0$, any (C^2, ε)-approximation f of g has no degenerate critical point in C.*

PROOF. We work in a coordinate neighborhood U_l. To put emphasis on the particular coordinate system $(x_1, x_2, \ldots, x_m)$ of U_l, we denote by

$$\left(\frac{\partial^2 g}{\partial x_i \partial x_j} \right)$$

the Hessian of g with respect to these coordinates. It is easily seen that there are no degenerate critical points of g in $C \cap K_l$ if and only if the condition

$$(2.54) \qquad \left| \frac{\partial g}{\partial x_1} \right| + \cdots + \left| \frac{\partial g}{\partial x_m} \right| + \left| \det \left(\frac{\partial^2 g}{\partial x_i \partial x_j} \right) \right| > 0$$

holds in $C \cap K_l$.

For a small enough $\varepsilon > 0$, the similar inequality

$$(2.55) \qquad \left| \frac{\partial f}{\partial x_1} \right| + \cdots + \left| \frac{\partial f}{\partial x_m} \right| + \left| \det \left(\frac{\partial^2 f}{\partial x_i \partial x_j} \right) \right| > 0$$

holds in $C \cap K_l$ for a (C^2, ε)-approximation f of g, as a consequence of (2.49). Hence f has no degenerate critical point in $C \cap K_l$. We repeat the above argument for $l = 1, 2, \ldots, k$ and conclude that f does not admit any degenerate critical point in $C \ (= \bigcup_{l=1}^{k} C \cap K_l)$. $\qquad \square$

Finally we are ready to prove the existence theorem (2.20) for Morse functions.

PROOF. Since we need (C^2, ε)-approximations of functions, we choose a covering of M by coordinate neighborhoods U_l ($l = 1, 2, \ldots, k$) as well as a covering of M by compact sets K_l in U_l ($l = 1, 2, \ldots, k$). We keep these coverings fixed throughout our proof.

We begin by taking the given function $g : M \to \mathbb{R}$ as f_0. The idea of the proof is that we construct functions f_l inductively with respect to l starting with f_0, so that f_l has no degenerate critical point in $K_1 \cup K_2 \cup \cdots \cup K_l$, $l = 1, 2, \ldots, k$. When we reach level $l = k$, we will have the desired Morse function f_k (because $M = K_1 \cup K_2 \cup \cdots \cup K_k$).

Denote by C_l the set $K_1 \cup K_2 \cup \cdots \cup K_l$ for simplicity. The C_l is compact since the union of a finite number of compact sets is compact. We formally set $C_0 = \emptyset$.

Suppose, as the inductive hypothesis, that we have constructed a function $f_{l-1} : M \to \mathbb{R}$ which has no degenerate critical point in C_{l-1}. We construct f_l using f_{l-1}.

We look at the coordinate neighborhood U_l and the compact set K_l in it. Let $(x_1, x_2, \ldots, x_m)$ be the coordinate system chosen for U_l. By Lemma 2.21, there exist real numbers $a_1, a_2, \ldots, a_m$ with small enough absolute values such that

$$(2.56) \qquad f_{l-1}(x_1, x_2, \ldots, x_m) - (a_1 x_1 + \cdots + a_m x_m)$$

is a Morse function on U_l. We must modify this function in such a way that the expression $a_1 x_1 + \cdots + a_m x_m$ makes sense outside the coordinate neighborhood U_l. To do this we use the following handy technique in manifold theory. The reader can find the proof in any textbook on manifold theory (for example, see [17]).

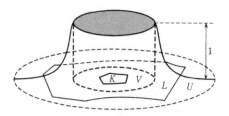

FIGURE 2.7. A step function h with respect to (U, K)

LEMMA 2.27. *Let U be a coordinate neighborhood and K a compact set in U. Then there exists a smooth function $h : U \to \mathbb{R}$ defined on U and satisfying the following properties (i), (ii), and (iii).*

(i) $0 \le h \le 1$.

(ii) h takes the value 1 on some open neighborhood V of K (V is an open set containing K).

(iii) h takes the value 0 outside some compact set L ($\subset U$) containing V.

(See Figure 2.7. We have $K \subset V \subset L \subset U$.)

We call such a function h a *step function* with respect to (U, K). The name "step function" may sound angular; however, h is a smooth function.

We apply the above lemma (2.27) to the coordinate neighborhood U_l and its compact set K_l to get a step function $h_l : U_l \to \mathbb{R}$ with respect to (U_l, K_l).

We construct the function f_l as follows (here L_l, which appears in the second case of the equation (2.57), is a compact set in U_l such that h_l constantly takes the value 0 outside of it):

(2.57)
$$
f_l = \begin{cases}
f_{l-1}(x_1, \ldots, x_m) - (a_1 m_1 + \cdots + a_m x_m)\, h_l(x_1, \ldots, x_m) \\
\hspace{7cm} \text{(in } U_l\text{)}, \\
f_{l-1}(x_1, \ldots, x_m) \hspace{4cm} \text{(outside } L_l\text{)}.
\end{cases}
$$

In the above equality, $h_l(x_1, \ldots, x_m) = 0$ on the intersection of the two sets (inside U_l and outside L_l, and this "outside" is open in

M), so that these two definitions agree on it to define a well-defined smooth function f_l on M.

The function f_l agrees with the "difference function" (2.56) in some neighborhood of the compact set K_l, where we have $h_l(x_1, \ldots, x_m) = 1$. Thus f_l is a Morse function on K_l, and so it has no degenerate critical points there.

Next we check that we can take f_l as a (C^2, ε)-approximation of f_{l-1}. We begin our calculation on U_l:

(2.58)
$$\begin{cases} |f_{l-1}(p) - f_l(p)| = |(a_1 x_1 + \cdots + a_m x_m)|h_l(p), \\ \left|\frac{\partial f_{l-1}}{\partial x_i}(p) - \frac{\partial f_l}{\partial x_i}(p)\right| = \left|a_i h_l(p) + (a_1 x_1 + \cdots + a_m x_m)\frac{\partial h_l}{\partial x_i}(p)\right|, \\ \qquad\qquad\qquad\qquad i = 1, 2 \ldots, m, \\ \left|\frac{\partial^2 f_{l-1}}{\partial x_i \partial x_j}(p) - \frac{\partial^2 f_l}{\partial x_i \partial x_j}(p)\right| \\ = \left|a_i \frac{\partial h_l}{\partial x_j}(p) + a_j \frac{\partial h_l}{\partial x_i}(p) + (a_1 x_1 + \cdots + a_m x_m)\frac{\partial^2 h_l}{\partial x_i \partial x_j}(p)\right|, \\ \qquad\qquad\qquad\qquad i, j = 1, 2, \ldots, m, \end{cases}$$

where $p = (x_1, x_2, \ldots, x_m)$.

Since the function h_l satisfies $0 \leq h_l \leq 1$ and is 0 outside the compact set L_l, the absolute value of its first and second order derivatives cannot exceed a certain positive number (the maximum–value theorem); therefore, we can make the right-hand side of (2.58) arbitrarily small by taking the absolute values of $a_1, \ldots, a_m$ small enough. In particular, f_l can be made (C^2, ε)-close to f_{l-1} in the compact set K_l.

To approximate f_l in the K_j's other than K_l, we use the coordinate system $(y_1, \ldots, y_m)$ of the coordinate neighborhood U_j of K_j and compute the differences between the respective first and second order derivatives of f_l and f_{l-1}. We evaluate these differences on K_j. To begin with, note that $f_l = f_{l-1}$ outside the compact set L_l, so that all we need to estimate is the difference on $K_j \cap L_l$, which is contained in the intersection $U_j \cap U_l$ of coordinate neighborhoods U_j and U_l. The difference on $U_j \cap U_l$ should be expressed by the right-hand side of (2.58) with a suitable Jacobi transformation between $(x_1, \ldots, x_m)$ and $(y_1, \ldots, y_m)$. By the maximum–value theorem, the absolute value of each component of the Jacobian on the compact set $K_j \cap L_l$ cannot exceed a certain value. Consequently, if we take the absolute values of $a_1, \ldots, a_m$ small enough, then the right-hand side of (2.58) on $K_j \cap L_l$ can be made as small as we wish. Thus, recalling that there is no difference between f_{l-1} and f_l outside L_l, we have shown that f_l is (C^2, ε)-close to f_{l-1}, as long as we take the absolute values of $a_1, \ldots, a_m$ sufficiently small.

Since we can repeat the above process for each $j = 1, 2, \ldots, k$, it follows by Definition 2.25 that we can indeed define f_l to be a (C^2, ε)-approximation of f_{l-1} for any positive $\varepsilon > 0$.

By the inductive hypothesis, f_{l-1} has no degenerate critical point in $C_{l-1} = K_1 \cup \cdots \cup K_{l-1}$. By Lemma 2.26, if f_l is (C^2, ε)-close to f_{l-1} for a sufficiently small $\varepsilon > 0$, it does not have any degenerate critical point in C_{l-1} either. Moreover, since we made sure that f_l has no degenerate critical point in K_l, it follows that f_l has no degenerate critical point in $C_{l-1} \cup K_l = C_l$.

Now we can proceed inductively for $l = 1, 2 \ldots, k$. The last f_k is a function which has no degenerate critical point in $C_k = M$; that is, f_k is a Morse function on M. Furthermore, by taking $\varepsilon > 0$ in each stage of the induction small enough, f_k can be made (C^2, ε')-close to the given function g for any specified ε'. This concludes the proof of the existence theorem 2.20 for Morse functions.   □

2.3. Gradient-like vector fields

Given a Morse function $f : M \to \mathbb{R}$, we consider a corresponding "gradient-like vector field," which plays an important role when we study how critical points of f are related with each other, and when we investigate handle decompositions of the manifold M. First we give a quick review of tangent vectors.

(a) Tangent vectors

Suppose that a manifold M is embedded in a Euclidian space $\mathbb{R}^N$ of a large enough dimension N. Let p be a point of M and consider a vector based at p which is tangent to M (see Figure 2.8). This is a *tangent vector of M at p*. The set of tangent vectors of M at p,

$$T_p(M),$$

forms a vector space called the *tangent vector space of M at p*. The tangent vector space of a manifold is a generalization of a tangent plane of a surface.

If the manifold M is m-dimensional, so is its tangent space $T_p(M)$ for $p \in M$.

A typical example of a tangent vector is a velocity vector of a curve. Let $c : (a, b) \to \mathbb{R}^N$ be a smooth curve in $\mathbb{R}^N$. Let

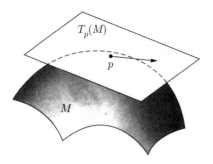

FIGURE 2.8. A tangent vector and the tangent vector space $T_p(M)$

$(X_1, X_2, \ldots, X_N)$ be the coordinates of $\mathbb{R}^N$ and let t be a parameter of c. Then c is described by

$$(2.59) \qquad c(t) = (X_1(t), X_2(t), \ldots, X_N(t)), \quad a < t < b.$$

For simplicity, we assume that the domain of the parameter (a, b) contains 0 and that the curve passes through the point p at $t = 0$: $c(0) = p$. The instantaneous velocity vector $\mathbf{v}$ of the curve at this moment $t = 0$ is given by

$$(2.60) \qquad \mathbf{v} = \frac{dc}{dt}(0) = \left(\frac{dX_1}{dt}(0), \frac{dX_2}{dt}(0), \ldots, \frac{dX_N}{dt}(0) \right).$$

If c is a curve in M, then this velocity vector $\dfrac{dc}{dt}(0)$ is a tangent vector of M at p (see Figure 2.9).

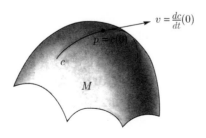

FIGURE 2.9. A velocity vector

Given a tangent vector, we can differentiate a function in its direction. Let us explain this using the coordinates $(X_1, X_2, \ldots, X_N)$ of $\mathbb{R}^N$ for now. Let $\mathbf{v} = (v_1, v_2, \ldots, v_N)$ be a tangent vector ($\in T_p(M)$) and let f be a function defined in a neighborhood of p in $\mathbb{R}^N$. We consider a curve $c(t) = (X_1, X_2, \ldots, X_N)$ in M which passes through p at $t = 0$. Suppose that the "initial velocity" (the velocity vector at $t = 0$) of this curve is $\mathbf{v}$:

$$(2.61) \qquad \frac{dc}{dt}(0) = \mathbf{v}; \text{ that is, } \frac{dX_j}{dt}(0) = v_j, \ j = 1, 2, \ldots, N.$$

If we consider the restriction of f to the curve c, we get a function $f(c(t))$ of one variable in t, which we differentiate at $t = 0$. Using the chain rule for the derivative of a composite function, we get

$$(2.62) \qquad \begin{aligned} \left. \frac{df(c(t))}{dt} \right|_{t=0} &= \left. \frac{d}{dt} f(X_1(t), X_2(t), \ldots, X_N(t)) \right|_{t=0} \\ &= \sum_{j=1}^{N} \frac{\partial f}{\partial X_j}(p) \frac{dX_j}{dt}(0) \\ &= \sum_{j=1}^{N} v_j \frac{\partial f}{\partial X_j}(p). \end{aligned}$$

The last line of this equation shows that the result depends only on f and $\mathbf{v}$, and does not depent on the curve c whose initial velocity is $\mathbf{v}$. Thus we can write this derivative as

$$(2.63) \qquad\qquad\qquad \mathbf{v} \cdot f,$$

which is the directional derivative of the function f in the direction $\mathbf{v}$.

From (2.62) we see that $\mathbf{v} \cdot f > 0$ if and only if the function $f(c(t))$ is an increasing function of t near $t = 0$, and hence $\mathbf{v} \cdot f > 0$ if and only if $\mathbf{v}$ points in the direction where f is increasing.

Next we consider a local coordinate sytem $(x_1, x_2, \ldots, x_m)$ about the point p in the manifold M, where $p = (a_1, a_2, \ldots, a_m)$. Consider a curve $c_i(t)$ which passes through the point p at time $t = 0$, and travels with unit velocity in the direction of the x_i-coordinate. In the local coordinate system in M, this curve is defined by

$$(2.64) \qquad c_i(t) = (a_1, \ldots, a_{i-1}, a_i + t, a_{i+1}, \ldots, a_m).$$

The velocity vector $\mathbf{e}_i$ of the curve $c_i(t)$ at $t = 0$ is the "basis vector" in the direction of x_i with respect to the local coordinate system

$(x_1, x_2, \ldots, x_m)$. We get the "basis vector" in each direction; then the set of "basis vectors"

$$(2.65) \qquad \mathbf{e}_1, \ \mathbf{e}_2, \ \ldots, \ \mathbf{e}_m$$

forms a basis of the tangent vector space $T_p(M)$ (see Figure 2.10).

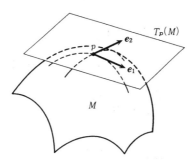

FIGURE 2.10. The vectors $\mathbf{e}_1, \ldots, \mathbf{e}_m$ form a basis of $T_p(M)$

The derivative of f in the direction of $\mathbf{e}_i$,

$$
\begin{aligned}
(2.66) \qquad \mathbf{e}_i \cdot f &= \frac{d}{dt} f(c_i(t)) \Big|_{t=0} \\
&= \frac{d}{dt} f(a_1, \ldots, a_{i-1}, a_i + t, a_{i+1}, \ldots, a_m) \Big|_{t=0} \\
&= \frac{\partial f}{\partial x_i}(p),
\end{aligned}
$$

agrees with the partial derivative with respect to x_i. For this reason, we often denote by

$$(2.67) \qquad \left(\frac{\partial}{\partial x_i} \right)_p$$

the basis vector $\mathbf{e}_i$ in the direction of x_i.

When two tangent vectors agree as directional differentiations, they are identical as tangent vectors; that is, if we have

$$(2.68) \qquad \mathbf{u} \cdot f = \mathbf{v} \cdot f$$

for any function f defined in a neighborhood of a point p, then $\mathbf{u} = \mathbf{v}$. We use this fact to derive a formula for the velocity vector $\mathbf{v}$ of $c(t)$ at $t = 0$ in terms of (as a linear combination of) the basis vectors $\mathbf{e}_1, \mathbf{e}_2,$ $\ldots, \mathbf{e}_m$, where $c(t)$ is a curve in M which passes through the point p

at $t = 0$ and $(x_1, x_2, \ldots, x_m)$ is a local coordinate system of M at p. Let

$$(2.69) \qquad c(t) = (x_1(t), x_2(t), \ldots, x_m(t)).$$

For a function f defined in some neighborhood of p we have

$$
\begin{aligned}
(2.70) \qquad \mathbf{v} \cdot f &= \frac{d}{dt} f(c(t)) \Big|_{t=0} \\
&= \frac{d}{dt} f(x_1(t), x_2(t), \ldots, x_m(t)) \Big|_{t=0} \\
&= \sum_{i=1}^{m} \frac{\partial f}{\partial x_i}(p) \frac{\partial x_i}{dt}(0) \\
&= \sum_{i=1}^{m} \frac{dx_i}{dt}(0) \, \mathbf{e}_i \cdot f,
\end{aligned}
$$

and hence we obtain the formula

$$(2.71) \qquad \mathbf{v} = \sum_{i=1}^{m} \frac{dx_i}{dt}(0) \, \mathbf{e}_i$$

for the velocity vector $\mathbf{v}$. If we use the expression $\left(\dfrac{\partial}{\partial x_i} \right)_p$ for $\mathbf{e}_i$, the above fomula becomes

$$(2.72) \qquad \mathbf{v} = \sum_{i=1}^{m} \frac{dx_i}{dt}(0) \left(\frac{\partial}{\partial x_i} \right)_p.$$

Similarly one can derive a formula for the coordinate transformation between two systems of basis vectors with respect to the coordinate systems $(x_1, \ldots, x_m)$ and $(y_1, \ldots, y_m)$:

$$(2.73) \qquad \left(\frac{\partial}{\partial x_i} \right)_p = \sum_{j=1}^{m} \frac{\partial y_j}{\partial x_i}(p) \left(\frac{\partial}{\partial y_j} \right)_p, \qquad i = 1, 2, \ldots, m.$$

So far we have assumed that M is embedded in $\mathbb{R}^N$; however, we can just as well discuss a more general situation to define the tangent space $T_p(M)$ as the vector space with the basis $\left(\dfrac{\partial}{\partial x_i} \right)_p$ ($i = 1, 2, \ldots, m$), and justify the coordinate transformation (2.73).

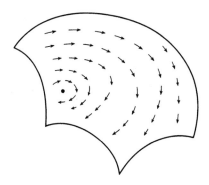

FIGURE 2.11. A vector field

(b) Vector fields

By a *vector field* on M, we mean a correspondence which assigns to each point p of M a tangent vector $\mathbf{v}$ at p (see Figure 2.11).

If U is a coordinate neighborhood in M with the coordinate system $(x_1, x_2, \ldots, x_m)$, a vector field X on U is described by

$$(2.74) \qquad X = \xi_1 \frac{\partial}{\partial x_1} + \xi_2 \frac{\partial}{\partial x_2} + \cdots + \xi_m \frac{\partial}{\partial x_m},$$

where $\xi_1, \xi_2, \ldots, \xi_m$ are functions defined on U. The above expression means that X is a vector field which assigns to each point p in U the tangent vector

$$\xi_1(p) \left(\frac{\partial}{\partial x_1} \right)_p + \xi_2(p) \left(\frac{\partial}{\partial x_2} \right)_p + \cdots + \xi_m(p) \left(\frac{\partial}{\partial x_m} \right)_p.$$

We say that X is a *smooth* vector field on U when every one of the functions $\xi_1, \xi_2, \ldots, \xi_m$ is smooth. We say that X is a *smooth* vector field on M if X is smooth on every coordinate neighborhood.

EXAMPLE 2.28. Let f be a function defined in a coordinate neighborhood U and let $(x_1, x_2, \ldots, x_m)$ be its coordinate system. Define a vector field X in U by

$$(2.75) \qquad X_f = \frac{\partial f}{\partial x_1} \frac{\partial}{\partial x_1} + \frac{\partial f}{\partial x_2} \frac{\partial}{\partial x_2} + \cdots + \frac{\partial f}{\partial x_m} \frac{\partial}{\partial x_m}.$$

This is the special case where we chose $\dfrac{\partial f}{\partial x_i}$ as the coefficient function ξ_i for each i.

We call X_f the *gradient vector field of the function* f.

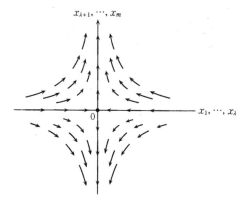

FIGURE 2.12. The gradient vector field of $f = -x_1^2 -$
$\cdots - x_\lambda^2 + x_{\lambda+1}^2 + \cdots + x_m^2$

A vector field itself is sort of a differential operator, since it assigns to each point a "tangent vector" which is a differential operation. Let us differentiate f with respect to the gradient vector field X_f:

$$
\begin{aligned}
X_f \cdot f &= \left(\sum_{i=1}^{m} \frac{\partial f}{\partial x_i} \frac{\partial}{\partial x_i} \right) \cdot f \\
&= \sum_{i=1}^{m} \left(\frac{\partial f}{\partial x_i} \right)^2 \\
&\geq 0.
\end{aligned}
$$
(2.76)

Notice that $(X_f \cdot f)(p) > 0$ unless p is a critical point of f:

$$
\frac{\partial f}{\partial x_1}(p) = \frac{\partial f}{\partial x_2}(p) = \cdots = \frac{\partial f}{\partial x_m}(p) = 0.
$$

In other words *the gradient vector field of f always points in a direction into which f is increasing, outside the critical points of f.*

We depict the gradient vector field of a Morse function in a standard form:

(2.77) $\qquad f = -x_1^2 - \cdots - x_\lambda^2 + x_{\lambda+1}^2 + \cdots + x_m^2$

in Figure 2.12 for $0 < \lambda < m$, and in Figure 2.13 for $\lambda = 0$ (the figure on the left) and for $\lambda = m$ (the figure on the right).

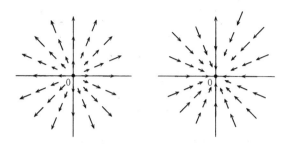

FIGURE 2.13. Gradient vector fields: $\lambda = 0$ (left) and $\lambda = m$ (right)

These gradient vector fields are written as

$$(2.78) \quad -2x_1\frac{\partial}{\partial x_1} - \cdots - 2x_\lambda\frac{\partial}{\partial x_\lambda} + 2x_{\lambda+1}\frac{\partial}{\partial x_{\lambda+1}} + \cdots + 2x_m\frac{\partial}{\partial x_m},$$

and will play an important role later.

So far the gradient vector field has been defined in a coordinate neighborhood U. Now we want to globalize.

(c) Gradient-like vector fields

Let f be a Morse function defined on a closed m-manifold. In the following discussion we always assume that X is a smooth vector field on M.

DEFINITION 2.29 (Gradient-like vector field). We say that X is a *gradient-like vector field* for a Morse function $f : M \to \mathbb{R}$ if the following two conditions hold.

(i) $X \cdot f > 0$ away from the critical points of f.

(ii) If p_0 is a critical point of f of index λ, then p_0 has a small enough neighborhood V with a suitable coordinate system $(x_1, x_2, \ldots, x_m)$ such that f has a standard form

$$(2.79) \qquad f = -x_1^2 - \cdots - x_\lambda^2 + x_{\lambda+1}^2 + \cdots + x_m^2 + f(p_0),$$

and X can be written as its gradient vector field:

$$(2.80)$$
$$X = -2x_1\frac{\partial}{\partial x_1} - \cdots - 2x_\lambda\frac{\partial}{\partial x_\lambda} + 2x_{\lambda+1}\frac{\partial}{\partial x_{\lambda+1}} + \cdots + 2x_m\frac{\partial}{\partial x_m}.$$

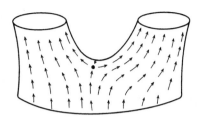

FIGURE 2.14. A gradient-like vector field

The first condition (i) says that outside the critical points of f, X points in the direction into which f is increasing. If we think of f as a height function, then X points "upward" (cf. Figure 2.14).

THEOREM 2.30. *Suppose that* $f : M \to \mathbb{R}$ *is a Morse function on a compact manifold* M. *Then there exists a gradient-like vector field* X *for* f.

PROOF. We cover M with a finite number of coordinate neighborhoods $U_1, U_2, \ldots, U_k$, as we did in the proof of the existence theorem 2.20 for Morse functions. We also choose a compact set K_j in each U_j in such a way that the compact sets $K_1, K_2, \ldots, K_k$ cover M as well. Moreover, we may assume that each critical point p_0 has a small neighborhood contained in exactly one coordinate neighborhood U_i and that f has a standard form in U_i (we first choose such a coordinate neighborhood for each critical point of f and then add enough U_j's to cover M).

We now construct for each $j = 1, 2, \ldots, k$ a gradient vector field X_j of f on U_j. We do this as in Example 2.28 by using the specified coordinate system $(x_1, x_2, \ldots, x_m)$ for U_j.

We have $X_j \cdot f > 0$ away from the critical points of f, so that X_j is pointed upward. Since we define each of these gradient vector fields using a specified coordinate system, there is no guarantee that X_i and X_j agree in the intersection $U_i \cap U_j$ of U_i and U_j if $i \neq j$. This presents a problem.

In order to define a smooth vector field X on all of M by putting the X_j together, the step functions of Lemma 2.27 become useful. Let

$$(2.81) \qquad\qquad h_j : U_j \to \mathbb{R}$$

be a step function corresponding to (U_j, K_j); h_j is a smooth function with $0 \leq h_j \leq 1$ such that $h_j = 1$ in some neighborhood V_j of K_j and $h_j = 0$ outside a compact set L_j containing V_j and contained in U_j. We extend h_j to a smooth function (using the same letter h_j) $h_j : M \to \mathbb{R}$ by setting $h_j = 0$ outside U_j.

Let us consider the vector field

$$(2.82) \qquad\qquad h_j X_j$$

which assigns to each point p in U_j "the vector $h_j(p)X_j(p)$, the scalar multiple of $X_j(p)$ by $h_j(p)$." We can extend this vector field to a smooth vector field on M by assigning the zero vector to each point outside U_j. We denote this vector field again by $h_j X_j$.

We construct such a vector field $h_j X_j$ in M for each $j = 1, 2, \ldots, k$, and consider their sum

$$(2.83) \qquad\qquad X = \sum_{j=1}^{k} h_j X_j.$$

We claim that X is a desired gradient-like vector field.

We first show that $X \cdot f > 0$ at a point p which is not a critical point. If U_j contains p, then $(X_j \cdot f)(p) > 0$, and if U_j does not contain p, then $h_j X_j(p) = \mathbf{0}$. Therefore, the derivative of f by each component of the above sum satisfies $(h_j X_j \cdot f)(p) \geq 0$. Recall, however, that the compact sets $K_1, K_2, \ldots, K_k$ cover M and hence p belongs to at least one of them, say, K_j, where $h_j = 1$ and $(X_j \cdot f)(p) > 0$. This shows that the derivative of f by at least one term in the sum (2.83) is surely positive, and $X \cdot f > 0$ is proved.

What about X at a critical point p_0? We have chosen a small neighborhood V of p_0 that is contained in one and only one U_i. In the neighborhood V, we have $h_j = 1$. Moreover, f is in a standard form on U_i, and hence $h_i X_i$ is a vector field of the form (2.80) in V. Since the rest of the terms $h_j X_j$ of X are $\mathbf{0}$ in V, we see that X satisfies the condition (ii) of Definition (2.29). This proves the existence of a gradient-like vector field. $\qquad\qquad \square$

The following gives a brief discussion on *integral curves* of a vector field X on a manifold M. We say that a curve $c(t)$ is an integral curve of a vector field X if we have

$$(2.84) \qquad\qquad \frac{dc}{dt}(t) = X_{c(t)}$$

for every t for which $c(t)$ is defined. Here $\dfrac{dc}{dt}(t)$ is the velocity vector of the curve c when the parameter value is t, and it is equal to the vector $X_{c(t)}$ which the vector field X specifies at the point $c(t)$. In other words, an integral curve of the vector field X is a flow line of a particle which moves with X as its velocity vectors. It is known that if M is a compact manifold without boundary, then there exists an integral curve $c_p(t)$ of X for $-\infty < t < \infty$ passing through p at $t = 0$ (cf. [11]).

If X is a gradient-like vector field of a Morse function $f : M \to \mathbb{R}$, then the integral curve $c_p(t)$ starting at an arbitrary point p approaches critical points as $t \to \infty$ and as $t \to -\infty$. As the curve approaches a critical point, the vectors of X becomes gradually smaller, and the speed of the curve gets slower, so that it never reaches the critical point. (On the other hand, if the curve starts at a critical point, then it can never leave this point, so that the integral curve is a singleton set).

Now let $f : M \to \mathbb{R}$ be a Morse function and let $[a, b]$ be a real interval. Set

(2.85) $$M_{[a,b]} = \{\, p \in M \mid a \leq f(p) \leq b \,\}.$$

Let us prove the following theorem, using a gradient-like vector field.

THEOREM 2.31. *If f has no critical value in the interval $[a, b]$, then $M_{[a,b]}$ is diffeomorphic to the product*

(2.86) $$f^{-1}(a) \times [0, 1]$$

(*cf. Figure 2.15*).

PROOF. Let X be a gradient-like vector field for f. As $X \cdot f > 0$ away from the critical points of f, we can define a new vector field Y on M with the critical points excluded (which is an open set) by

$$Y = \frac{1}{X \cdot f} X.$$

Since by hypothesis $M_{[a,b]}$ contains no critical points of f, it is in the domain of the vector field Y (the vector field Y runs over it). Consider the integral curve $c_p(t)$ of Y which starts at a point p of

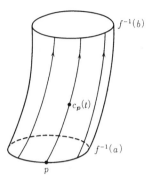

FIGURE 2.15. If there is no critical point, then $M_{[a,b]}$ is diffeomorphic to the product $f^{-1}(a) \times [0,1]$

$f^{-1}(a)$. Using the definition of the velocity vector, we obtain

$$
\begin{aligned}
\frac{d}{dt} f(c_p(t)) &= \frac{dc}{dt}(t) \cdot f \\
&= Y_{c(t)} \cdot f \\
&= \frac{1}{X \cdot f} X \cdot f \\
&= 1.
\end{aligned}
$$

Thus the integral curve $c_p(t)$ maintains an upward climb with the constant speed 1 with respect to the "height" defined by f. Since it starts at the level $f = a$ at the time $t = 0$, it will reach the level $f = b$ at the time $t = b - a$. Define a map $h : f^{-1}(a) \times [0, b-a] \to M_{[a,b]}$ by

$$
h(p,t) = c_p(t).
$$

We can show that h is a diffeomorphism by using the facts that $c_p(t)$ depends smoothly on both p and t, and that two distinct integral curves do not meet (the uniqueness of integral curves) (cf. [11]). We have therefore proved that $M_{[a,b]} \cong f^{-1}(a) \times [0, b-a]$, and with the observation that $f^{-1}(a) \times [0, b-a] \cong f^{-1}(a) \times [0,1]$, we complete the proof of the theorem. $\square$

We can adopt the technique used in the proof of Theorem 2.30 or Theorem 2.31 for a function $f : M \to \mathbb{R}$ which is not necessarily Morse. The next theorem on the existence of a bicollar neighborhood is such an example.

Let $f : M \to \mathbb{R}$ be a smooth function (not necessarily Morse) such that 0 is not a critical value. By the implicit function theorem (Theorem 2.3), $f^{-1}(0)$ is a submanifold of M, which we denote by K. If M is m-dimensional, K is $(m-1)$-dimensional. We then have the following theorem.

THEOREM 2.32. (Existence of a bicollar neighborhood). *Suppose that $f^{-1}(0) = K$ is compact. Then for a sufficiently small number $\varepsilon > 0$, there exist an open neighborhood U ($U \subset M$) of K and a diffeomorphism $h : K \times (-\varepsilon, \varepsilon) \to U$ satisfying the following conditions:*

(i) $h(K, 0) = K$ (K corresponds to $K \times 0$).

(ii) $f(h(K, t)) = t$, $-\varepsilon < t < \varepsilon$ (the "level surface" of $f = t$ corresponds to $K \times t$).

We call an open neighborhood (such as U) of K having the product structure $K \times (-\varepsilon, \varepsilon)$ a *bicollar neighborhood* of K.

Let us prove Theorem 2.32.

PROOF. Since $K = f^{-1}(0)$ is compact, $f : M \to \mathbb{R}$ has no critical values in $[-\varepsilon, \varepsilon]$ for a small enough positive number ε. If we set $U = f^{-1}(-\varepsilon, \varepsilon)$, then we can think of the restriction of f to U as a Morse function on U (having no critical points). Let X be a gradient-like vector field for f on U and set $Y = 1/(X \cdot f)X$. Now we only need to define a diffeomorphism

$$h : K \times (-\varepsilon, \varepsilon) \to U$$

by $h(p, t) = c_p(t)$, where $c_p(t)$ is the integral curve of Y starting at p of K.   □

We have the following corollary to the above theorem, which shows the existence of a collar neighborhood of the boundary for manifolds with boundary.

COROLLARY 2.33 (Existence of a collar neighborhood). *Let N be a manifold with boundary. If the boundary ∂N is compact, then there exists a neighborhood V of ∂N with a diffeomorphism $h : \partial N \times [0, 1) \to V$. Here $[0, 1)$ is a half-open interval. Note also that $h(\partial N, 0) = \partial N$ (the boundary ∂N corresponds to $\partial N \times 0$).*

We say that a neighborhood of the boundary with the product structure, such as V above, is a *collar neighborhood* of the boundary. See Figure 2.16. The collar neighborhood of Corollary 2.33 is an open set; however, sometimes one defines a collar neighborhood as the image $h(\partial N \times [0, a])$ of $\partial N \times [0, a]$ by h, where a is any real number,

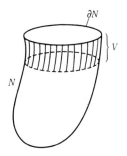

FIGURE 2.16. A collar neighborhood

$0 < a < 1$. According to such a definition, a collar neighborhood is a closed set in N.

We give a brief proof of the corollary.

PROOF. We may assume that the manifold N with boundary is of the form $M_{f \geq 0}$ for some manifold M without boundary and some smooth function $f : M \to \mathbb{R}$ such that 0 is not its critical value. The boundary ∂N of N corresponds to $f^{-1}(0)$ and it is compact by assumption. By Theorem 2.32, $f^{-1}(0)$ has a bicollar neighborhood U in M having the product structure with a diffeomorphism $h : f^{-1}(0) \times (-\varepsilon, \varepsilon) \to U$. For the collar neighborhood V we seek in Corollary 2.33, we can simply take the half of U, $h(f^{-1}(0) \times [0, \varepsilon))$, and for the diffeomorphism we take the composite

$$\partial N \times [0, 1) \xrightarrow{\cong} f^{-1}(0) \times [0, \varepsilon) \xrightarrow{h|f^{-1}(0) \times [0,\varepsilon)} V.$$

$\square$

2.4. Raising and lowering critical points

In this section we discuss the simplest case of raising and lowering critical points. We will deal with a more dynamic version of this method in Chapter 3.

THEOREM 2.34. *Let $f : M \to \mathbb{R}$ be a Morse function on M, and let $p_1, p_2, \ldots, p_r$ be its critical points. Then there exists a Morse function f' whose critical points are $p_1, p_2, \ldots, p_r$ such that*

$$f'(p_i) \neq f'(p_j) \ \text{if} \ p_i \neq p_j, \quad i, j = 1, 2, \ldots, r.$$

Moreover, we can take f' as (C^2, ε)-close to f as we wish.

PROOF. We assume that the critical value of f at the critical points p_1 and p_2 is the same c, and try to modify f slightly. By Theorem 2.16 we may choose a local coordinate system $(x_1, x_2, \ldots, x_m)$ about p_1 and write f in a standard form:

$$f = -x_1^2 - \cdots - x_\lambda^2 + x_{\lambda+1}^2 + \cdots + x_m^2 + c.$$

Let X_f be a gradient vector field for f with respect to this coordinate system, and calculate $X_f \cdot f$:

$$X_f \cdot f = \left(\frac{\partial f}{\partial x_1}\right)^2 + \cdots + \left(\frac{\partial f}{\partial x_\lambda}\right)^2 + \cdots + \left(\frac{\partial f}{\partial x_m}\right)^2$$
$$= 4(x_1^2 + \cdots + x_\lambda^2 + \cdots + x_m^2).$$

For a sufficiently small number $\varepsilon > 0$, consider the m-disks D_ε and $D_{2\varepsilon}$ of radii ε and 2ε, respectively, both centered at p_1. It follows from the above equality that $4\varepsilon^2 \leq X_f \cdot f \leq 4(2\varepsilon)^2$ in the region $D_{2\varepsilon} - \operatorname{int} D_\varepsilon$.

Denote by K the compact set D_ε and by U the interior $\operatorname{int}(D_{2\varepsilon})$ of $D_{2\varepsilon}$, so that U is an open set containing a compact set K. Consider a step function $h : U \to \mathbb{R}$ with respect to (U, K) (cf. Lemma 2.27). We extend h to a smooth function on the entire manifold M by setting $h = 0$ outside U. We denote this new function again by h. Define a function $\tilde{f}$ by

$$\tilde{f} = f + ah,$$

where a is a small enough non-zero real number. Since $f = \tilde{f}$ outside U, f and $\tilde{f}$ have the same set of critical points there. Since $h = 1$ in $\operatorname{int}(D_\varepsilon)$, we also see that the point p_1 is the only critical point of both f and $\tilde{f}$ in this region. Hence, the only place left where $\tilde{f}$ possibly has a different set of critical points from f is the region between D_ε and $D_{2\varepsilon}$. We compute the differences of the first partial derivatives of f and $\tilde{f}$:

$$\left|\frac{\partial f}{\partial x_i} - \frac{\partial \tilde{f}}{\partial x_i}\right| = \left|a\frac{\partial h}{\partial x_i}\right|, \quad i = 1, 2, \ldots, m.$$

It follows, therefore, that as long as the absolute value of a is small enough, we can make the difference between $\sum_{i=1}^{m} \left(\frac{\partial f}{\partial x_i}\right)^2$ and $\sum_{i=1}^{m} \left(\frac{\partial \tilde{f}}{\partial x_i}\right)^2$ arbitrarily small. As we observed earlier, $\sum_{i=1}^{m} \left(\frac{\partial f}{\partial x_i}\right)^2$

$(= X_f \cdot f)$ takes the minimum value $4\varepsilon^2 > 0$ between $D_{2\varepsilon}$ and D_ε, and hence, if a is small enough, then $\displaystyle\sum_{i=1}^{m} \left(\frac{\partial \tilde{f}}{\partial x_i} \right)^2$ also takes a non-zero minimum value there. Thus $\tilde{f}$ cannot have critical points between the disks D_ε and $D_{2\varepsilon}$. We conclude that f and $\tilde{f}$ have the same set of critical points. Noting that a non-degenerate critical point of f is also non-degenerate for $\tilde{f}$, we see that $\tilde{f}$ is a Morse function. Furthermore, we have

$$\tilde{f}(p_1) = f(p_1) + a, \quad \tilde{f}(p_2) = f(p_2);$$

therefore, even though $f(p_1) = f(p_2)$, we have $\tilde{f}(p_1) \neq \tilde{f}(p_2)$.

We continue the above argument to define a function whose critical values are distinct for distinct critical points.

To prove that $\tilde{f}$ is (C^2, ε)-close to f, we only need to follow the argument in the proof of Theorem 2.20 (existence of Morse functions). □

Summary
2.1 Non-degenerate and degenerate critical points are defined for a function on an m-dimensional manifold using the Hessian of the function.
2.2 A function can be represented by a standard form in some neighborhood of a non-degenerate critical point.
2.3 A function with only non-degenerate critical points is called a Morse function.
2.4 There exists a Morse function on any compact manifold M.
2.5 There exists a gradient-like vector field for a given Morse function defined on any compact manifold.
2.6 A given Morse function on a compact manifold can be modified without changing its critical points in such a way that the critical points of the new function take distinct critical values.

Exercises
2.1 Show that the definition of a critical point p_0 of a function $f : M \to \mathbb{R}$ does not depend on the choice of a local coordinate system about p_0.
2.2 Let $S^{m-1} = \{ (x_1, \ldots, x_m) \,|\, x_1^2 + \cdots + x_m^2 = 1 \}$ be the unit $(m-1)$-sphere, and define the "height function" $f : S^{m-1} \to \mathbb{R}$ by $f(x_1, \ldots, x_m) = x_m$. Prove that f is a Morse function on S^{m-1}, and

determine its critical points and their respective indices (the coordinates $(x_1, \ldots, x_m)$ are not a local coordinate system of S^{m-1}! You may assume that $m \geq 2$).

2.3 Let $f : M \to \mathbb{R}$ and $N \to \mathbb{R}$ be Morse functions, where M and N are closed manifolds. Define a function $F : M \times N \to \mathbb{R}$ by

$$F = (A + f)(B + g),$$

where A and B are positive numbers. Show that F is a Morse function if A and B are sufficiently large, and determine its critical points and their respective indices in terms of those of f and g.

2.4 We look at the *m-dimensional torus* $T^m = S^1 \times S^1 \times \cdots \times S^1$, the product of m circles. We write a point of T^m as $(\theta_1, \theta_2, \ldots, \theta_m)$ where each θ_i represents the angle of the i-th circle component of the point, and define $f : T^m \to \mathbb{R}$ by

$$f(\theta_1, \theta_2, \ldots, \theta_m) = (R + \cos\theta_1)(R + \cos\theta_2) \cdots (R + \cos\theta_m),$$

where $R > 1$. Show that f is a Morse function, and find its critical points and their respective indices.

CHAPTER 3

Handlebodies

In Chapter 1 we expressed closed surfaces as the union of some collection of handles of indices 0, 1 and 2. In Chapter 2 we described the theory of Morse functions on general manifolds. In this chapter we use the results of Chapter 2 to investigate handlebody decompositions of compact manifolds. We provide a general theory of handlebodies in Section 1, and give some examples in Section 2. In Sections 3 and 4 we remold handlebodies by "sliding" and "canceling" handles.

3.1. Handle decompositions of manifolds

Let M be a closed manifold and $f : M \to \mathbb{R}$ a Morse function on M. Set

$$(3.1) \qquad M_t = \{\, p \in M \mid f(p) \leq t \,\},$$

for a value t of f. We investigate how M_t changes as the parameter t changes, as we did for closed surfaces in Chapter 1.

THEOREM 3.1. *If f has no critical values in the real interval $[a, b]$, then M_a and M_b are diffeomorphic: $M_a \cong M_b$.*

We use Theorem 2.31 for the proof of this theorem, which is exactly the same as the result (Lemma 1.23) for closed surfaces. Geometrically what we do is to let the manifold M_a "flow" along a gradient-like (upward) vector field X (after adjusting the speed of the flow by multiplying X by the function $\dfrac{1}{X \cdot f}$). Then, after a certain period of time, M_a meets and coincides with M_b. We leave the details of the proof to the reader.

The problem we investigate, therefore, is the change in shape of M_t as the parameter t passes through a critical value. According to Theorem 2.34, we may assume that f takes distinct critical values at distinct critical points. We also notice that f has only a finite number, say $n + 1$, of critical points. We order the critical values of f in ascending order and label the corresponding critical points as

(3.2) $p_0, \ p_1, \ \ldots, \ p_n.$

In this chapter we always start the subscripts of the critical points with zero, for later convenience.

Setting $c_i = f(p_i)$, we have

(3.3) $c_0 < c_1 < \cdots < c_n,$

where c_0 is the minimum value and c_n is the maximum value.

It is evident here that no point p with $f(p) < c_0$ exists, so that $M_t = \emptyset$ if $t < c_0$. Similarly, $M_t = M$ if $c_n \leq t$, since $f(p) \leq c_n$ for all points p of M.

We now pursue the changes of M_t around the minimum and maximum values of f.

We start with the minimum value. In our case we are assuming that p_0 is the only point which gives the minimum value. We write f in a standard form,

(3.4) $f = x_1^2 + x_2^2 + \cdots + x_m^2 + c_0.$

Since c_0 is the minimum value of f, the values of f cannot be less than c_0, and hence the standard form (3.4) has no negative signs in the quadratic part. That is, the index of p_0 is necessarily 0.

Let $\varepsilon > 0$ be a sufficiently small positive number. As we saw above, $M_{c_0 - \varepsilon} = \emptyset$, but we may use the standard form (3.4) to express $M_{c_0 + \varepsilon}$ as

(3.5) $M_{c_0 + \varepsilon} = \left\{ (x_1, \ldots, x_m) \mid x_1^2 + \cdots + x_m^2 \leq \varepsilon \right\};$

that is, $M_{c_0 + \varepsilon}$ is diffeomorphic to the m-dimensional disk D^m. The standard form of f shows that it takes the minimum value c_0 at the center of the disk $(x_1, \ldots, x_m) = (0, \ldots, 0)$. The function f takes values which increase as f approaches the boundary of the disk, and it attains the value $c_0 + \varepsilon$ on the boundary of the disk. Thus, we would like to draw this m-dimensional disk as a bowl whose rim points upward, as we did in Chapter 1 in the case $m = 2$. The space we live in is 3-dimensional, however, so it is rather difficult to do so for general dimension m. Figure 3.1 shows $M_{c_0 + \varepsilon}$ for $m = 3$ (the 3-dimensional disk = an ordinary solid ball). We urge the reader to imagine this 3-dimensional disk curving "upward" in the fourth dimension.

So far we have dealt with the situation around the minimum value, but even if c_i is not the minimum value, each time t passes

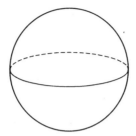

FIGURE 3.1. An m-dimensional 0-handle, where $m = 3$

through a critical value c_i of index 0 we add an m-dimensional disk facing upward, and $M_{c_i+\varepsilon}$ becomes diffeomorphic to $M_{c_i-\varepsilon} \sqcup D^m$ (disjoint union). The m-dimensional (upward) disk which appears at a critical point of index 0 is called a 0-*handle*, or more precisely, an m-dimensional 0-handle.

We now consider the situation around the maximum value c_n. We are assuming that there is only one point p_n where f attains the maximum value. Since c_n is the maximum value, f cannot take values larger than c_n, so that the quadratic part of the following standard form for f has no positive signs:

$$(3.6) \qquad f = -x_1^2 - x_2^2 - \cdots - x_m^2 + c_n.$$

Thus the index of p_n is necessarily m.

When $c_n < t$ we have $M_t = M$. We can express the manifold $M_{c_n-\varepsilon}$ just before the value of t reaches c_n as

$$(3.7) \qquad x_1^2 + x_2^2 + \cdots + x_m^2 \geq \varepsilon,$$

by taking the right-hand side of the standard form (3.6) to be $\leq c_n - \varepsilon$. This manifold corresponds to the complement of the m-dimensional disk D^m of radius $\sqrt{\varepsilon}$. The boundary $\partial M_{c_n-\varepsilon}$ is the boundary S^{m-1} of this m-dimensional disk of radius $\sqrt{\varepsilon}$.

As t increases from $c_n - \varepsilon$ and passes c_n, the boundary of $M_{c_n-\varepsilon}$ is capped off with a lid which is the m-dimensional disk. Thus we see the completion of a compact m-manifold without boundary. The function f takes the maximum value at the center of the disk. The value of f decreases toward the boundary of the disk, and so the m-disk attached here points "downward." We call this m-disk an m-handle – more precisely, an m-dimensional m-handle. In general, when t passes any critical value c_i with index m, the m-handle caps a connected component of the boundary of $M_{c_i - \varepsilon}$. This boundary is an $(m-1)$-sphere S^{m-1}, and $M_{c_i - \varepsilon}$ is the manifold just before t reaches c_i.

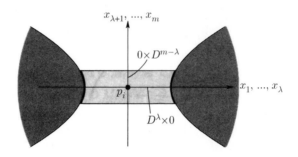

FIGURE 3.2. A λ-handle

We now have a complete picture around critical points of indices 0 and m. Next we consider the changes of M_t near a critical point p_i of a general index λ $(0 < \lambda < m)$ as t passes the corresponding critical value c_i.

We take a local coordinate system about the critical point p_i of index λ and put f in a standard form:

$$(3.8) \qquad f = -x_1^2 - \cdots - x_\lambda^2 + x_{\lambda+1}^2 + \cdots + x_m^2 + c_i.$$

The situation around p_i will be clarified in Theorem 3.2. The situation of Figure 3.2 is approximately that of Theorem 3.2. Around p_i in Figure 3.2, the darkly shaded area in the figure depicts $M_{c_i - \varepsilon}$, which we get by setting $f \le c_i - \varepsilon$ in (3.8); that is,

$$(3.9) \qquad x_1^2 + \cdots + x_\lambda^2 - x_{\lambda+1}^2 - \cdots - x_m^2 \ge \varepsilon.$$

The lightly shaded area corresponds to the inequalities

$$(3.10) \qquad \begin{cases} x_1^2 + \cdots + x_\lambda^2 - x_{\lambda+1}^2 - \cdots - x_m^2 \le \varepsilon, \\ x_{\lambda+1}^2 + \cdots + x_m^2 \le \delta, \end{cases}$$

where δ is a positive number much smaller than ε, and this lightly shaded area is called an m-dimensional handle of index λ, or more briefly an m-dimensional λ-*handle*. The reader should check that this handle is diffeomorphic to the direct product of the λ-disk and the

$(m - \lambda)$-disk

$$D^\lambda \times D^{m-\lambda}.$$

The λ-disk

(3.11) $D^\lambda \times \mathbf{0} = \{\, (x_1, \ldots, x_\lambda, 0, \ldots, 0) \mid x_1^2 + \cdots + +x_\lambda^2 \leq \varepsilon \,\}$

is the *core* of the λ-handle, and the $(m - \lambda)$-disk

(3.12)
$$\mathbf{0} \times D^{m-\lambda} = \{\, (0, \ldots, 0, x_{\lambda+1}, \ldots, x_m) \mid x_{\lambda+1}^2 + \cdots + x_m^2 \leq \delta \,\}$$

intersecting the core is the *co-core*. (The core and the co-core intersect *transversely*, that is, orthogonally in some coordinate system. Compare also Definition 3.29.) The name co-core means that it is dual to the core and indicates the thickness of the λ-handle. The core and the co-core intersect transversely at the origin $(0,0)$, which is exactly the critical point p_i.

We attach a λ-handle $D^\lambda \times D^{m-\lambda}$ to $M_{c_i-\varepsilon}$ as in Figure 3.2:

(3.13) $M_{c_i-\varepsilon} \cup D^\lambda \times D^{m-\lambda}.$

The next theorem describes the change of M_t as the parameter t passes through the critical value c_i of index λ.

THEOREM 3.2. *The set $M_{c_i+\varepsilon}$ is diffeomorphic to the manifold obtained by attaching a λ-handle to $M_{c_i-\varepsilon}$:*

(3.14) $M_{c_i+\varepsilon} \cong M_{c_i-\varepsilon} \cup D^\lambda \times D^{m-\lambda}.$

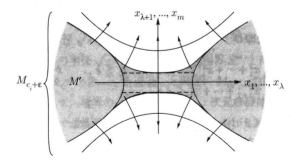

FIGURE 3.3. The smoothed-out manifold M' after attaching a λ-handle to $M_{c_i-\varepsilon}$

The space $M_{c_i-\varepsilon}$ with a λ-handle attached is not "smooth" at the "corners" of the boundary where the handle meets $M_{c_i-\varepsilon}$. We must "smooth" these corners out to make a C^∞-manifold M' as shown in Figure 3.3. (Cf. [12] or Remark 1.3.3 in [3].) A more accurate statement of Theorem 3.2, therefore, should be that $M_{c_i+\varepsilon}$ is diffeomorphic to the smoothed-out manifold M'.

Notice that in Figure 3.3, $M_{c_i+\varepsilon}$ corresponds to the piece defined by

$$(3.15) \qquad x_1^2 + \cdots x_\lambda^2 - x_{\lambda+1}^2 - \cdots - x_m^2 \geq -\varepsilon.$$

The proof of Theorem 3.2 is just about the same as the proof of Theorem 3.1. The idea of the proof is as follows. We again use the gradient-like (upward) vector field X for f. One can see in Figure 3.3 that the vector field X, after leaving the boundary $\partial M'$ of M', continues to flow upward till it reaches the boundary $\partial M_{c_i+\varepsilon}$ of $M_{c_i+\varepsilon}$. We may multiply the vector field X by a suitable function, and let M' "flow" along X so that it will match $M_{c_i+\varepsilon}$ after a certain period of time. This will show that M' and $M_{c_i+\varepsilon}$ are diffeomorphic.

We will not worry much about the non-smooth edges of the boundary of $M_{c_i-\varepsilon} \cup D^\lambda \times D^{m-\lambda}$, and we will pretend that it is in fact a smoothed-out manifold M' in our discussion.

We now go back to the standard form (3.8) of f and trace the change of the values of f on the core D^λ of a λ-handle. At the center of the core f takes the critical value c_i, it is decreasing as it approaches the boundary of the disk, and on the boundary f takes the value $c_i - \varepsilon$. The core D^λ, therefore, is an "upside-down" λ-disk. The function f on the co-core $D^{m-\lambda}$ attains the critical value c_i at the center $(0,0)$, and its value increases as it approaches the boundary of the disk, where it takes the value $c_i + \delta$. The co-core, therefore, is an "upright" disk; the core faces down and the co-core faces up, like a horse saddle or a mountain pass.

The core of a 0-handle is 0-dimensional and the co-core is m-dimensional, so that there is no downward direction; every direction points upward. On the other hand, the core of an m-handle is m-dimensional and the co-core is 0-dimensional so that any m-handle "faces down."

One attaches a λ-handle $D^\lambda \times D^{m-\lambda}$ to $M_{c_i-\varepsilon}$ by pasting $\partial D^\lambda \times D^{m-\lambda}$ along the boundary $\partial M_{c_i-\varepsilon}$ of $M_{c_i-\varepsilon}$ (cf. Figure 3.2). In order to describe the handle-attaching accurately, one must specify a map

$$(3.16) \qquad \varphi : \partial D^\lambda \times D^{m-\lambda} \to \partial M_{c_i-\varepsilon},$$

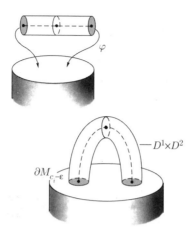

FIGURE 3.4. A 1-handle

indicating where each point of $\partial D^\lambda \times D^{m-\lambda}$ corresponds to in the boundary $\partial M_{c_i - \varepsilon}$. We identify each $p \in \partial D^\lambda \times D^{m-\lambda}$ with the point $\varphi(p) \in \partial M_{c_i - \varepsilon}$.

The map φ is a smooth "embedding" which we call the *attaching map* of the λ-handle. The boundary ∂D^λ of the core disk is a $(\lambda - 1)$-dimensional sphere $S^{\lambda - 1}$, and is called the *attaching sphere*. An attaching map is an embedding

$$(3.17) \qquad \varphi : S^{\lambda - 1} \times D^{m - \lambda} \to \partial M_{c_i - \varepsilon}$$

of a thickened $(\lambda - 1)$-sphere $S^{\lambda - 1} \times D^{m - \lambda}$ (with the $(m - \lambda)$-dimensional thickness) into the boundary $\partial M_{c_i - \varepsilon}$.

In Figures 3.4 and 3.5, a 3-dimensional 1-handle and a 2-handle are depicted, respectively. The picture of a 1-handle justifies the name "handle." The 2-handle is depicted as a thickened upside-down bowl.

DEFINITION 3.3 (Handlebody [23]). A manifold (with boundary in general) obtained from D^m by attaching handles of various indices one after another

$$(3.18) \qquad D^m \cup D^{\lambda_1} \times D^{m - \lambda_1} \cup \cdots \cup D^{\lambda_n} \times D^{m - \lambda_n}$$

is called an m-dimensional *handlebody*.

More precisely, a handlebody is defined in three steps as follows.

(i) A disk D^m is an m-dimensional handlebody.

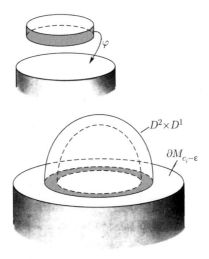

FIGURE 3.5. A 2-handle

(ii) The manifold

$$D^m \cup_{\varphi_1} D^{\lambda_1} \times D^{m-\lambda_1}$$

obtained from D^m by attaching a λ_1-handle with an attaching map
of class C^∞, $\varphi_1 : \partial D^{\lambda_1} \times D^{m-\lambda_1} \to \partial D^m$, is an m-dimensional handlebody, denoted by $\mathcal{H}(D^m; \varphi_1)$.

(iii) If $N = \mathcal{H}(D^m; \varphi_1, \cdots, \varphi_{i-1})$ is an m-dimensional handlebody, then the manifold

$$N \cup_{\varphi_i} D^{\lambda_i} \times D^{m-\lambda_i}$$

obtained from N by attaching a λ_i-handle $D^{\lambda_i} \times D^{m-\lambda_i}$ with an
attaching map of class C^∞, $\varphi_i : \partial D^{\lambda_i} \times D^{m-\lambda_i} \to \partial N$, is an m-
dimensional handlebody, denoted by $\mathcal{H}(D^m; \varphi_1, \cdots, \varphi_{i-1}, \varphi_i)$.

In fact, the manifold is smoothed out each time a handle is
attached, so that the resulting handlebodies are always considered
smooth manifolds [12].

The attachment of a 0-handle is nothing but a disjoint union, so
that in the case of index $\lambda_i = 0$, there is no need of specifying the
attaching map φ_i. Therefore, in the case $\lambda_i = 0$, the attaching map
φ_i in the notation $\mathcal{H}(D^m; \varphi_1, \cdots, \varphi_{i-1}, \varphi_i)$ does not have a meaning,
but the notation φ_i is kept formally. Also, in the above definition,

although it is logically sufficient to have only the conditions (i) and (iii), the condition (ii) is added to help in visualizing a geometric image of a handlebody.

THEOREM 3.4 (Handle decomposition of a manifold). *When a Morse function $f : M \to \mathbb{R}$ is given on a closed manifold M, a structure of a handlebody on M is determined by f. The handles of this handlebody correspond to the critical points of f, and the indices of the handles coincide with the indices of the corresponding critical points.*

In other words, M can be expressed as a handlebody. When a manifold is expressed as a handlebody, it is called a *handle decomposition*.

PROOF. We may assume that all the critical points of the given Morse function $f : M \to \mathbb{R}$ have distinct critical values. Arrange the critical points in such a way that their critical values are in ascending order, and name them

$$p_0, \; p_1, \; \cdots, \; p_n.$$

Let λ_i be the index of the critical point p_i. Fix a gradient-like vector field X on M for f.

Theorem 3.4 is proved by induction on the subscripts i of the critical points p_i. Let c_i be the value of f at p_i, and we will show that $M_{c_i+\varepsilon}$ is a handlebody.

First for $i = 0$, the index of the critical point p_0 is 0, as p_0 gives the minimum value of f. Therefore, $M_{c_0+\varepsilon}$ is diffeomorphic to the m-dimensional disk D^m. From the first step (i) of the definition of a handlebody, D^m is indeed an m-dimensional handlebody, and the statement of the induction is proved for $i = 0$. In this case $M_{c_0+\varepsilon}$ is a 0-handle itself.

Next we make the inductive assumption that $M_{c_{i-1}+\varepsilon}$ is a handlebody $\mathcal{H}(D^m; \varphi_1, \cdots, \varphi_{i-1})$, and will prove that $M_{c_i+\varepsilon}$ is a handlebody $\mathcal{H}(D^m; \varphi_1, \cdots, \varphi_{i-1}, \varphi_i)$.

Recall that $M_{c_i+\varepsilon}$ is diffeomorphic to a manifold obtained from $M_{c_{i-1}+\varepsilon}$ by attaching a λ_i-handle (Theorem 3.2). The attaching map

(3.19) $\qquad \varphi : \partial D^{\lambda_i} \times D^{m-\lambda_i} \to \partial M_{c_i-\varepsilon}$

of this handle is, as shown in Figure 3.2, determined naturally without ambiguity. The interval $[c_{i-1} + \varepsilon, c_i - \varepsilon]$ contains no critical values, so that $M_{c_{i-1}+\varepsilon}$ is diffeomorphic to $M_{c_i-\varepsilon}$ (Theorem 3.1). This diffeomorphism is given by letting $M_{c_{i-1}+\varepsilon}$ flow along the gradient-like vector field X until it matches $M_{c_i-\varepsilon}$. Let

$$(3.20) \qquad\qquad \Phi : M_{c_{i-1}+\varepsilon} \to M_{c_i-\varepsilon}$$

be such a diffeomorphism.

From the induction hypothesis that $M_{c_{i-1}+\varepsilon}$ is a handlebody $\mathcal{H}(D^m; \varphi_1, \cdots, \varphi_{i-1})$, we see that $M_{c_i-\varepsilon}$ is diffeomorphic to the same handlebody. Therefore the manifold $M_{c_i+\varepsilon}$, which is obtained from $M_{c_i-\varepsilon}$ by attaching a λ_i-handle, is also diffeomorphic to a handlebody (step (iii) of the three-step definition).

This completes the proof of Theorem 3.4. However, let us investigate the attaching map of the new λ_i-handle in some more detail. From the induction hypothesis, $M_{c_{i-1}+\varepsilon}$ is the handlebody

$$\mathcal{H}(D^m; \varphi_1, \cdots, \varphi_{i-1}),$$

so that the above diffeomorphism Φ can be regarded as a diffeomorphism from the handlebody $\mathcal{H}(D^m; \varphi_1, \cdots, \varphi_{i-1})$ to $M_{c_i-\varepsilon}$. Precisely speaking, the λ_i-handle is not attached directly to the handlebody $\mathcal{H}(D^m; \varphi_1, \cdots, \varphi_{i-1})$, but rather is attached to $M_{c_i-\varepsilon}$, and its attaching map, φ in (3.19), is naturally determined. Therefore, when $M_{c_i-\varepsilon}$ is identified with the handlebody $\mathcal{H}(D^m; \varphi_1, \cdots, \varphi_{i-1})$ by the diffeomorphism Φ in (3.20), the λ_i-handle is attached to the handlebody $\mathcal{H}(D^m; \varphi_1, \cdots, \varphi_{i-1})$ by the composition map

$$\Phi^{-1} \circ \varphi : \partial D^{\lambda_i} \times D^{m-\lambda_i} \to \partial\{\mathcal{H}(D^m; \varphi_1, \cdots, \varphi_{i-1})\}$$

where Φ^{-1} is the inverse map of Φ. If we denote this attaching map $\Phi^{-1} \circ \varphi$ by φ_i, then $M_{c_i+\varepsilon}$ is the handlebody

$$\mathcal{H}(D^m; \varphi_1, \cdots, \varphi_{i-1}) \cup_{\varphi_i} D^{\lambda_i} \times D^{m-\lambda_i} = \mathcal{H}(D^m; \varphi_1, \cdots, \varphi_{i-1}, \varphi_i).$$

This completes the proof of Theorem 3.4.　　　　　　　　□

REMARK. The above proof implies the following. When a handle decomposition of a manifold M is obtained from a Morse function $f : M \to \mathbb{R}$,

(i) the order of the handles and their indices are determined by the critical points of f, and

(ii) the attaching maps φ_i ($= \Phi^{-1} \circ \varphi$) of the handles are determined by a gradient-like vector field X for f (since Φ is determined by X).

Therefore, even with the same f, the structure of a handle decomposition may vary if one changes the choice of a gradient-like vector field X for f (as the attaching map may vary). This fact will become important in Section 3.3.

3.2. Examples

In this section we will look at Morse functions on some manifolds.

EXAMPLE 3.5 (The m-dimensional sphere). Let

$$(3.21) \qquad S^m = \{\, (x_1, \cdots, x_m, x_{m+1}) \mid x_1^2 + \cdots x_m^2 + x_{m+1}^2 = 1 \,\}$$

be the m-dimensional sphere, and define a function $f : S^m \to \mathbb{R}$ by

$$(3.22) \qquad\qquad f(x_1, \cdots, x_m, x_{m+1}) = x_{m+1}.$$

Then f is the height fuction with respect to the $(m+1)$-th coordinate. It is easy to see that f is a Morse function. There are only two critical points of f, $(0, \cdots, 0, -1)$ and $(0, \cdots, 0, 1)$, and their indices are 0 and m, respectively (Exercise 2.2). Therefore, by Theorem 3.4, S^m has a handle decomposition consisting of one 0-handle and one m-handle:

$$(3.23) \qquad\qquad S^m = D^m \cup D^m.$$

This corresponds to the decomposition of the sphere into the southern hemisphere and the northern hemisphere.

Conversely, the following theorem holds.

THEOREM 3.6. *If there is a Morse function* $f : M \to \mathbb{R}$ *on an* m-*dimensional closed manifold* M *with only two critical points, then* M *is homeomorphic to* S^m. *Furthermore, if* $m \leq 6$, *then* M *is diffeomorphic to* S^m.

Theorem 1.16 is the case $m = 2$ in the above theorem. In that case, it was concluded that a closed surface M is diffeomorphic to the 2-dimensional sphere S^2, under the condition that there is a Morse function with two critical points. The same conclusion holds when the dimension is less than or equal to 6, but for general dimension m larger than or equal to 7, it can be concluded only that M is homeomorphic to S^m as in the above theorem, and not that it is diffeomorphic. In fact, in most of the dimensions higher than or equal to 7, there exist smooth m-dimensional manifolds M which are homeomorphic to the m-dimensional sphere S^m but are not diffeomorphic to it. Such manifolds are called *exotic spheres*. Exotic spheres were discovered first by J. Milnor in 1956 in dimension 7 [13].

The proof of Theorem 3.6 is similar to that of Theorem 1.16 in Chapter 1. As one sees in the proof, the step that fails when one tries to conclude that M is diffeomorphic to S^m in general dimension m is to extend a diffeomorphism between $(m-1)$-dimensional spheres

$$(3.24) \qquad\qquad h : S^{m-1} \to S^{m-1}$$

to a diffeomorphism between m-dimensional disks

$$(3.25) \qquad\qquad \tilde{h} : D^m \to D^m.$$

Such extensions exist for $m \leq 6$ (for $m = 2$ see the answer to Exercise 1.3 of this book, for $m = 3$ see [22], for $m = 4$ see [1], and for $m = 5, 6$ this is a consequence of the disk theorem [18] and the uniqueness of the differential structure of the 5- and 6-spheres, see [5] and [14]), but may not exist for $m \geq 7$. However, if one does not require the differentiability of $\tilde{h}$, then in any dimension, any homeomorphism $h :$ $S^{m-1} \to S^{m-1}$ extends to a homeomorphism $\tilde{h} : D^m \to D^m$ (Exercise 3.1). This is the reason why Theorem 3.6 holds if the conclusion is about homeomorphisms instead of diffeomorphisms.

EXAMPLE 3.7. (The direct product $S^m \times S^n$ of spheres). Let $f_m : S^m \to \mathbb{R}$ and $f_n : S^n \to \mathbb{R}$ be the height functions in Example 3.5. Let A and B be real numbers such that $1 < A < B$. Then

$$(3.26) \qquad f = (A + f_m)(B + f_n) : S^m \times S^n \to \mathbb{R}$$

is a Morse function on the product space $S^m \times S^n$. There are four critical points of f, and their indices are 0, n, m, and $m + n$ (cf. Exercise 2.3). The critical values are, respectively, $(A - 1)(B - 1)$, $(A - 1)(B + 1)$, $(A + 1)(B - 1)$, and $(A + 1)(B + 1)$. (Note that this is also in ascending order of the real numbers.)

Therefore, $S^m \times S^n$ is obtained from D^{m+n} by attaching an n-handle, an m-handle, and an $(m + n)$-handle of dimension $(m + n)$, one after another:

$$(3.27) \qquad S^m \times S^n = D^{m+n} \cup D^n \times D^m \cup D^m \times D^n \cup D^{m+n}.$$

EXAMPLE 3.8 (Projective space P^m). Consider the set of all lines through the origin $\mathbf{0}$ in the $(m + 1)$-dimensional Euclidian space $\mathbb{R}^{m+1}$. The *projective space* P^m is this set with a structure of an m-dimensional manifold. For any point $(x_1, \cdots, x_m, x_{m+1})$ other than the origin $\mathbf{0}$, a line that passes through the points $(x_1, \cdots, x_m, x_{m+1})$ and $\mathbf{0}$ is uniquely determined. Since this line is a "point" of P^m, it is considered that the point $(x_1, \cdots, x_m, x_{m+1})$ determines a single point of P^m. This point of P^m is denoted by $[x_1, \cdots, x_m, x_{m+1}]$. A necessary and sufficient condition for two lines to coincide in $\mathbb{R}^{m+1}$, one through the point $(y_1, \cdots, y_m, y_{m+1})$ and the origin $\mathbf{0}$, and the other through the point $(x_1, \cdots, x_m, x_{m+1})$ and the origin $\mathbf{0}$, is that there exists a non-zero real number α such that

$$(3.28) \qquad (y_1, \cdots, y_m, y_{m+1}) = (\alpha x_1, \cdots, \alpha x_m, \alpha x_{m+1}).$$

Therefore, the condition (3.28) is a necessary and sufficient condition for two corresponding points in P^m to coincide, that is,

$$[y_1, \cdots, y_m, y_{m+1}] = [x_1, \cdots, x_m, x_{m+1}].$$

Using this condition, one proves that P^m is compact as follows. For any point $[x_1, \cdots, x_m, x_{m+1}]$ of P^m, one can choose α in (3.28) such that $(y_1, \cdots, y_m, y_{m+1})$ satisfies

$$(3.29) \qquad y_1^2 + \cdots + y_m^2 + y_{m+1}^2 = 1.$$

With this condition, $(y_1, \cdots, y_m, y_{m+1})$ is a point of the unit sphere S^m in $\mathbb{R}^{m+1}$, and furthermore, in P^m, $[y_1, \cdots, y_m, y_{m+1}]$ is the same point as $[x_1, \cdots, x_m, x_{m+1}]$. Therefore the mapping

$$S^m \to P^m$$

which assigns $[y_1, \cdots, y_m, y_{m+1}]$ to $(y_1, \cdots, y_m, y_{m+1})$ is an "onto" continuous mapping, since any given $[x_1, \cdots, x_m, x_{m+1}]$ is assigned to the point $(y_1, \cdots, y_m, y_{m+1})$ of the unit sphere. Being a bounded closed subset of $\mathbb{R}^{m+1}$, S^m is compact; hence its continuous image P^m is also compact. This proves that P^m is compact. (The image of a compact set under a continuous map is compact.)

The above map $S^m \to P^m$ is called the *projection*. The projection is a 2-to-1 mapping which assigns the same point of P^m to two points $(y_1, \cdots, y_m, y_{m+1})$ and $(-y_1, \cdots, -y_m, -y_{m+1})$ of S^m.

Define a function $f : P^m \to \mathbb{R}$ by

$$(3.30) \qquad \begin{aligned} &f([x_1, \cdots, x_m, x_{m+1}]) \\ &= \frac{a_1 x_1^2 + \cdots + a_m x_m^2 + a_{m+1} x_{m+1}^2}{x_1^2 + \cdots + x_m^2 + x_{m+1}^2}, \end{aligned}$$

where $a_1, \cdots, a_m, a_{m+1}$ are arbitrarily fixed real constants satisfying $a_1 < \cdots < a_m < a_{m+1}$. The value of the function remains unchanged after mutiplying all the x_i's simultaneously by α, so this function is indeed defined as a function on P^m.

Fix a subscript i and consider the set U_i consisting of points $[x_1, \cdots, x_m, x_{m+1}]$ of P^m with $x_i \neq 0$; then U_i is an open set of P^m.

There is an m-dimensional local coordinate system $(X_1, \cdots, X_m)$ on U_i defined as follows:

(3.31)
$$X_1 = \frac{x_1}{x_i}, \quad \cdots, \quad X_{i-1} = \frac{x_{i-1}}{x_i}, \quad X_i = \frac{x_{i+1}}{x_i}, \quad \cdots, \quad X_m = \frac{x_{m+1}}{x_i}.$$

(Notice that the subscripts of X and x shift at the i-th coordinate.)

By dividing the numerator and the denominator in the right-hand side of the defining relation (3.30) of f by x_i^2, we obtain an expression representing f in terms of the local coordinate system $(X_1, \cdots, X_m)$:

(3.32)
$$f(X_1, \cdots, X_m)$$
$$= \frac{a_1 X_1^2 + \cdots + a_{i-1} X_{i-1}^2 + a_i + a_{i+1} X_i^2 + \cdots + a_{m+1} X_m^2}{X_1^2 + \cdots + X_{i-1}^2 + 1 + X_i^2 + \cdots + X_m^2}.$$

To find the critical values, we obtain

(3.33)
$$\frac{\partial f}{\partial X_m}$$
$$= \frac{2X_m\{(a_{m+1} - a_1)X_1^2 + \cdots + (a_{m+1} - a_m)X_{m-1}^2 + (a_{m+1} - a_i)\}}{(X_1^2 + \cdots + X_m^2 + 1)^2},$$

by differentiating f with respect to X_m. Since a_{m+1} is larger than the other a_j's ($j = 1, \cdots, m$), the right-hand side of the above expression is 0 if and only if $X_m = 0$.

Next consider the restriction $f|_{X_m=0}$ of f on $X_m = 0$:

(3.34)
$$f|_{X_m=0}(X_1, \cdots, X_{m-1})$$
$$= \frac{a_1 X_1^2 + \cdots + a_{i-1} X_{i-1}^2 + a_i + a_{i+1} X_i^2 + \cdots + a_m X_{m-1}^2}{X_1^2 + \cdots + X_{i-1}^2 + 1 + X_i^2 + \cdots + X_{m-1}^2}.$$

By differentiating $f|_{X_m=0}$ with respect to X_{m-1}, we see that the derivative is 0 if and only if $X_{m-1} = 0$, for the same reason as above. Continuing this process, we see that the critical points of f on the coordinate neighborhood U_i must satisfy

$$X_i = \cdots = X_{m-1} = X_m = 0.$$

Next differentiate (3.32) with respect to X_1 and use the fact that a_1 is smaller than $a_2, \cdots, a_{m+1}$ to see that $\dfrac{\partial f}{\partial X_1}$ is 0 if and only if $X_1 = 0$. Furthermore, differentiate $f|_{X_1=0}$ with respect to X_2 to see that the derivative is 0 if and only if $X_2 = 0$. Repeating this argument, we have that the critial points of f on U_i must satisfy

$$X_1 = X_2 = \cdots = X_{i-1} = 0.$$

In summary, the only critical point of f on U_i is the origin $(0, \cdots, 0)$ of the local coordinate system $(X_1, \cdots, X_m)$. In the notation $[x_1, \cdots, x_m, x_{m+1}]$, the point $[0, \cdots, 0, 1, 0, \cdots, 0]$ is the only critical point in U_i, where the entry 1 is at the i-th coordinate. Check that the Hessian $\left(\dfrac{\partial^2 f}{\partial X_j \partial X_k} \right)$ of f at this critical point is

(3.35)
$$\begin{pmatrix} 2(a_1 - a_i) & & & & & \\ & \ddots & & & & \\ & & 2(a_{i-1} - a_i) & & & \\ & & & 2(a_{i+1} - a_i) & & \\ & & & & \ddots & \\ & & & & & 2(a_{m+1} - a_i) \end{pmatrix}$$

where all the entries other than diagonal ones are 0's. The determinant of the Hessian is not 0, since the diagonal entries up to and including the $(i-1)$-th entry are negative and the others are positive, as $a_1 < \cdots < a_i < \cdots < a_{m+1}$. Therefore the critical point at the origin of U_i is non-degenerate, and has index $i - 1$. Also, the value of the function at this point is a_i.

Since P^m is covered by $(m + 1)$ coordinate neighborhoods U_i $(i = 1, \cdots, m + 1)$, we have proved the following: the Morse function $f : P^m \to \mathbb{R}$ we constructed here has $(m + 1)$ critical points whose indices are

$$0, \ 1, \ \cdots, \ m$$

in ascending order. Therefore P^m has a handle decomposition

(3.36)
$$P^m = D^m \cup D^1 \times D^{m-1} \cup D^2 \times D^{m-2} \cup \cdots \cup D^{m-1} \times D^1 \cup D^m.$$

In particular, the 1-dimensional projective space $P^1 = D^1 \cup D^1$ is diffeomorphic to the circle S^1.

EXAMPLE 3.9 (Handle decomposition of P^2). The 2-dimensional projective space P^2 is called the *projective plane*. We analyze the handle decomposition of the projective plane in some detail. According to Example 3.8, P^2 is a 2-dimensional closed manifold, and its handle decomposition consists of a 0-handle, a 1-handle, and a 2-handle attached in this order. There are two ways to attach a 1-handle to a 0-handle (a disk). They are depicted in Figure 3.6 left and right. If it is attached as in the left of the figure, then an annulus results, and it is impossible to obtain a closed surface by capping it off with only one 2-handle. Therefore the situation must be as in the right of the figure in the case of the projective plane. In the right, the union of a 0-handle and a 1-handle is homeomorphic to a Möbius band, whose boundary is a single circle, so that it can be capped off with a single 2-handle. (Of course, this is possible only abstractly, and is impossible in our 3-dimensional space.)

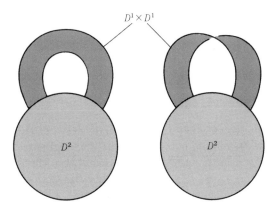

$D^1 \times D^1$

D^2 D^2

FIGURE 3.6. Two ways to attach a 2-dimensional 1-handle

In summary, a handle decomposition of the projective plane P^2 is obtained by attaching a 1-handle on a disk as in Figure 3.6 right, and then by attaching a 2-handle. It can also be said that it is obtained from a Möbius band and a disk by pasting them together along their boundaries.

The reader may wonder whether there are other ways to attach a 1-handle to a 0-handle, for example as in Figure 3.7. The way depicted in Figure 3.7, however, is the same as the left of Figure 3.6, and the manifolds thus obtained in both cases are homeomorphic to an annulus. The difference between Figure 3.6 left and Figure 3.7 is the way the annulus is embedded in 3-dimensional space $\mathbb{R}^3$.

We will discuss the handle decomposition of the 3-dimensional projective space P^3 in Chapter 5.

EXAMPLE 3.10 (Complex projective space $\mathbb{C}P^m$). The set of all $(m + 1)$-tuples of complex numbers

$$(z_1, \cdots, z_m, z_{m+1})$$

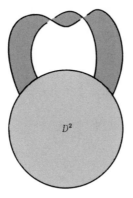

FIGURE 3.7. This way of attaching a handle is the same as the left of Figure 3.6

is denoted by $\mathbb{C}^{m+1}$ and is called the $(m+1)$-dimensional complex space. The m-dimensional *complex projective space* is the set of all "lines in the sense of complex numbers" in the space $\mathbb{C}^{m+1}$ that pass through the origin $\mathbf{0}$, with a structure of a complex manifold of complex dimension m, and is denoted by $\mathbb{C}P^m$. Here the superscript m represents the fact that the complex dimension of $\mathbb{C}P^m$ is m, but the dimension of $\mathbb{C}P^m$ in the ordinary sense is $2m$. This manifold is also known to be compact. The projective space P^m in Example 3.8 is sometimes called the *real projective space* to distinguish it from $\mathbb{C}P^m$.

A point $(z_1, \cdots, z_m, z_{m+1})$ of $\mathbb{C}^{m+1}$ distinct from the origin determines a point of $\mathbb{C}P^m$. This point is denoted by $[z_1, \cdots, z_m, z_{m+1}]$. A necessary and sufficient condition for $[z_1, \cdots, z_m, z_{m+1}] = [w_1, \cdots, w_m, w_{m+1}]$ is that there exists a non-zero complex number α such that

$$(3.37) \qquad (z_1, \cdots, z_m, z_{m+1}) = (\alpha w_1, \cdots, \alpha w_m, \alpha w_{m+1}).$$

For real numbers $a_1 < \cdots < a_m < a_{m+1}$, define a function $f : \mathbb{C}P^m \to \mathbb{R}$ by

$$(3.38) \qquad \begin{aligned} &f(z_1, \cdots, z_m, z_{m+1}) \\ &= \frac{a_1|z_1|^2 + \cdots + a_m|z_m|^2 + a_{m+1}|z_{m+1}|^2}{|z_1|^2 + \cdots + |z_m|^2 + |z_{m+1}|^2}. \end{aligned}$$

In a similar way as in the case of real projective spaces, fix a subscript i and define the set U_i of all points $[z_1, \cdots, z_m, z_{m+1}]$ with $z_i \neq 0$. Then U_i is an open set of $\mathbb{C}P^m$.

On this set there is an m-dimensional complex local coordinate system defined by

(3.39) $Z_1 = \dfrac{z_1}{z_i}, \cdots, Z_{i-1} = \dfrac{z_{i-1}}{z_i}, Z_i = \dfrac{z_{i+1}}{z_i}, \cdots, Z_m = \dfrac{z_{m+1}}{z_i}.$

Using $(Z_1, \cdots, Z_m)$, f can be written as

(3.40)

$$ f = \frac{a_1|Z_1|^2 + \cdots + a_{i-1}|Z_{i-1}|^2 + a_i + a_{i+1}|Z_i|^2 + \cdots + a_{m+1}|Z_m|^2}{|Z_1|^2 + \cdots + |Z_{i-1}|^2 + 1 + |Z_i|^2 + \cdots + |Z_m|^2}. $$

If we set $Z_j = X_j + \sqrt{-1}\, Y_j$ $(j = 1, \cdots, m)$, then we have

(3.41) $|Z_j|^2 = X_j^2 + Y_j^2.$

By regarding f as a function of $2m$ variables $(X_1, Y_1, \cdots, X_m, Y_m)$, and using similar arguments as in the case of P^m, we find that the origin $(0, \cdots, 0)$ is the only critical point in U_i, that it is non-degenerate, and that its index is $2(i-1)$ (the critical value is a_i).

Since the space $\mathbb{C}P^m$ can be covered by $(m+1)$ complex coordinate neighborhoods U_i $(i = 1, \cdots, m, m+1)$, we have proved the following: the Morse function $f : \mathbb{C}P^m \to \mathbb{R}$ has $m+1$ critical points, and their indices are

$$ 0, \ 2, \ \cdots, \ 2m, $$

in ascending order. Therefore we obtain the following handle decomposition of $\mathbb{C}P^m$:

(3.42) $\mathbb{C}P^m = D^{2m} \cup D^2 \times D^{2m-2} \cup \cdots \cup D^{2m-2} \times D^2 \cup D^{2m}.$

In particular, the complex 1-dimensional projective space $\mathbb{C}P^1$ is called the *complex projective line*. It is a 2-dimensional compact manifold, and there exists a Morse function with only two critical points. Therefore, by Theorem 3.6, $\mathbb{C}P^1$ is diffeomorphic to the 2-dimensional sphere S^2.

Also, the complex 2-dimensional projective space $\mathbb{C}P^2$ is called the *complex projective plane*. The complex projective plane is a typical 4-dimensional manifold, and will be investigated in some more detail in Chapter 5.

EXAMPLE 3.11 (Rotation group $SO(m)$). First we review orthogonal matrices. An $m \times m$ real matrix $A = (a_{ij})$ is called an *orthogonal matrix* if it satisfies the following two conditions.

(i) The i-th column $\mathbf{a}_i$ of A, regarded as a column vector, has unit length. That is, for any $i = 1, \cdots, m$,

(3.43) $$|\mathbf{a}_i|^2 = a_{1i}^2 + a_{2i}^2 + \cdots + a_{mi}^2 = 1.$$

(ii) If $i \neq k$, then $\mathbf{a}_i$ and $\mathbf{a}_k$ are orthogonal. That is, for $i, k = 1, \cdots, m$, $i \neq k$, the inner product $\mathbf{a}_i \cdot \mathbf{a}_k$ of $\mathbf{a}_i$ and $\mathbf{a}_k$ is 0:

(3.44) $$\mathbf{a}_i \cdot \mathbf{a}_k = a_{1i} \cdot a_{1k} + a_{2i} \cdot a_{2k} + \cdots + a_{mi} \cdot a_{mk} = 0.$$

These two properties are equivalent to the condition ${}^t\!A\, A = E$, where ${}^t\!A$ denotes the transpose matrix of A, and E denotes the $m \times m$ identity matrix. Considering the determinants, we obtain $\det A = \pm 1$.

An orthogonal matrix with $\det A = 1$ is called a *rotation matrix*, and we denote by $SO(m)$ the set of all $m \times m$ rotation matrices, and call it the *rotation group* of degree m. The set $SO(m)$ forms a group under matrix product. Furthermore, $SO(m)$ is a smooth manifold, and the group multiplication is of class C^∞ with respect to the manifold structure. Such a group is called a *Lie group*. The rotation group $SO(m)$ is a typical example of a Lie group.

The group $SO(1)$ is the trivial group, so we assume from now on that $m \geq 2$, and investigate the manifold structure of $SO(m)$. From the definition of an orthogonal group, $SO(m)$ is a closed subset of the set of all arrangements of m column vectors of length 1 ($\cong S^{m-1} \times \cdots \times S^{m-1}$), so that $SO(m)$ is compact.

Next we compute their dimensions. For the first column vector $\mathbf{a}_1$, we can choose an arbitrary vector in $\mathbb{R}^m$ of unit length, so that the choice of the first column vector contributes $(m-1)$ dimensions. Once $\mathbf{a}_1$ is chosen, the second vector $\mathbf{a}_2$ lies in the $(m-1)$-dimensional Euclidean space which is orthogonal to $\mathbf{a}_1$, and we can choose an arbitrary vector of unit length in this space as $\mathbf{a}_2$, so that the choice of $\mathbf{a}_2$ contributes $(m-2)$ dimensions. Repeating this process, we compute that the dimension of $SO(m)$ is

$$(m-1) + (m-2) + \cdots + 2 + 1 = \frac{m(m-1)}{2}.$$

For any fixed real numbers $1 < c_1 < c_2 < \cdots < c_m$, define a function $f : SO(m) \to \mathbb{R}$ by

(3.45) $$f(A) = c_1 x_{11} + c_2 x_{22} + \cdots + c_m x_{mm},$$

where

(3.46)
$$A = \begin{pmatrix} x_{11} & x_{12} & \cdots & x_{1m} \\ x_{21} & x_{22} & \cdots & x_{2m} \\ & \cdots & \cdots & \\ x_{m1} & x_{m2} & \cdots & x_{mm} \end{pmatrix}.$$

LEMMA 3.12. *The critical points of the function* $f : SO(m) \to \mathbb{R}$ *defined in (3.45) are*

(3.47)
$$\begin{pmatrix} \pm 1 & 0 & \cdots & 0 \\ 0 & \pm 1 & \cdots & 0 \\ & & \ddots & \\ 0 & 0 & \cdots & \pm 1 \end{pmatrix}.$$

(The combinations of signs of the diagonal entries can be chosen arbitrarily as long as the determinant is 1.)

PROOF. We introduce a specific rotation matrix B_θ. The matrix B_θ is defined by

(3.48)
$$\begin{pmatrix} \cos\theta & -\sin\theta & 0 & \cdots & 0 \\ \sin\theta & \cos\theta & 0 & \cdots & 0 \\ 0 & 0 & 1 & \cdots & 0 \\ & & & \ddots & \\ 0 & 0 & 0 & \cdots & 1 \end{pmatrix}.$$

That is, the matrix B_θ is the rotation by the angle θ in the plane in $\mathbb{R}^m$ spanned by the first and the second coordinate axes.

The product AB_θ of an orthogonal matrix A given generally by the expression (3.46) and the matrix B_θ defined above forms a curve in $SO(m)$ with parameter θ. Furthermore, this curve passes through A at $\theta = 0$. By actually computing the matrix product and substituting it into the definition of the function f, we obtain

(3.49)
$$f(AB_\theta) = c_1(x_{11}\cos\theta + x_{12}\sin\theta) + c_2(-x_{21}\sin\theta + x_{22}\cos\theta)$$
$$+ c_3 x_{33} + \cdots + c_m x_{mm}.$$

By differentiating this with respect to θ, and evaluating at $\theta = 0$ (i.e., by differentiating the function f in the direction of the velocity vector

$$\left.\frac{d}{d\theta}AB_\theta\right|_{\theta=0}$$ of the curve AB_θ at the "point" A), we obtain

$$(3.50) \qquad \left.\frac{d}{d\theta}f(AB_\theta)\right|_{\theta=0} = c_1 x_{12} - c_2 x_{21}.$$

If we compute the derivative for a similar curve $B_\theta A$, we obtain

$$(3.51) \qquad \left.\frac{d}{d\theta}f(B_\theta A)\right|_{\theta=0} = -c_1 x_{21} + c_2 x_{12}.$$

If a point A is a critical point of $f : SO(m) \to \mathbb{R}$, then the derivatives (3.50) and (3.51) at A must both be 0. From the assumption that $1 < c_1 < c_2$, however, it follows that

$$(3.52) \qquad x_{12} = x_{21} = 0,$$

if the right-hand sides of (3.50) and (3.51) vanish.

The above conclusion was obtained by using the rotation matrix B_θ of the plane in $\mathbb{R}^m$ spanned by the first and the second coordinate axes, but if we use the same argument using the rotation matrix of the plane spanned by the i-th and the k-th coordinate axes ($1 \leq i < k \leq m$), then we obtain

$$(3.53) \qquad x_{ik} = x_{ki} = 0, \quad \forall i, k, \quad 1 \leq i < k \leq m,$$

provided A is a critical point. That is, A is a diagonal matrix (which has non-zero entries only on the diagonal). Since A is a rotation matrix, A must have the form stated in Lemma 3.12.

Conversely, to show that the matrices of the form stated in Lemma 3.12 are indeed critical points of the function $f : SO(m) \to \mathbb{R}$, regard the space of all $m \times m$ matrices as the m^2-dimensional Euclidean space $\mathbb{R}^{m^2}$, and consider $SO(m)$ as being embedded in it. Let A be one of the matrices described in the lemma, that is, a matrix with diagonal entries $\varepsilon_1, \varepsilon_2, \cdots, \varepsilon_m$ ($\varepsilon_j = \pm 1$) and all the other entries 0. When we compute the velocity vector of the curve AB_θ (at the point A) evaluated at $\theta = 0$, a simple calculation shows that

$$(3.54) \qquad \left.\frac{d}{d\theta}AB_\theta\right|_{\theta=0} = A\left.\frac{d}{d\theta}B_\theta\right|_{\theta=0} = \begin{pmatrix} 0 & -\varepsilon_1 & 0 & \cdots & 0 \\ \varepsilon_2 & 0 & 0 & \cdots & 0 \\ 0 & 0 & 0 & \cdots & 0 \\ & & & \ddots & \\ 0 & 0 & 0 & \cdots & 0 \end{pmatrix}.$$

Denote the matrix of the right-hand side by V_{12}. Although it may be hard to imagine, here the matrix V_{12} is regarded as a vector in the m^2-dimensional Euclidean space $\mathbb{R}^{m^2}$.

In this case we used the velocity vector of the curve AB_θ defined with the rotation matrix B_θ of the plane in $\mathbb{R}^m$ spanned by the first and the second axes, but if we compute the velocity vector V_{ik} at A of the curve similarly defined using the rotation matrix of the plane spanned by the i-th and k-th axes, we obtain a matrix with the (i,k)-entry $-\varepsilon_i$, (k,i)-entry ε_k, and all the other entries 0. When we consider all i and k with $1 \leq i < k \leq m$, these vectors V_{ik} are linearly independent in $\mathbb{R}^{m^2}$, and the number of such vectors is the same as the number of ways of choosing (i,k) with $i < k$, i.e., ${}_mC_2 = \dfrac{m(m-1)}{2}$. This equals the dimension of $SO(m)$, so we see that the vectors V_{ik} $(1 \leq i < k \leq m)$ form a basis of the tangent vector space $T_A(SO(m))$ of $SO(m)$ at A.

If A is a matrix of the form given in Lemma 3.12, then it can be computed that the derivative of f in the direction of any V_{ik} is 0, in a similar way as in (3.50). This proves that A is a critical point of f. $\qquad\qquad\qquad\qquad\qquad\qquad\qquad\qquad\qquad\qquad\qquad\qquad\quad\Box$

Let A be a matrix of the form stated in Lemma 3.12 ($\varepsilon_1, \varepsilon_2, \cdots,$ ε_m on the diagonal and 0 for the other entries), and we compute the Hessian H_f of f at A. For this purpose, a local coordinate system of $SO(m)$ near A must be specified.

Denote by $B_\theta^{(ik)}$ the rotation matrix of the plane in $\mathbb{R}^m$ spanned by the i-th and k-th coordinate axes. There are $\dfrac{m(m-1)}{2}$ of them, when we consider all such matrices over all choices of i and k $(i < k)$. A local coordinate system $(\theta, \varphi, \psi, \cdots)$ of dimension $\dfrac{m(m-1)}{2}$ is given by the matrices obtained by multiplying the matrix A by all these matrices from the right in an arbitrarily fixed order

$$(3.55) \qquad\qquad A \cdot B_\theta^{(ik)} B_\varphi^{(jl)} B_\psi^{(hr)} \cdots ,$$

which enables us to compute the Hessian. By actually computing the derivative at A, we obtain

$$\frac{\partial^2 f}{\partial\theta\partial\varphi}(A \cdot B_\theta^{(ik)} B_\varphi^{(jl)})\Big|_{(\theta,\varphi)=(0,0)}$$

(3.56)
$$= f\left(A \cdot \frac{d}{d\theta}B_\theta^{(ik)}\Big|_{\theta=0} \cdot \frac{d}{d\varphi}B_\varphi^{(jl)}\Big|_{\varphi=0}\right)$$

$$= \begin{cases} 0 & (i,k) \neq (j,l), \\ -c_i\varepsilon_i - c_k\varepsilon_k & (i,k) = (j,l). \end{cases}$$

Here, we may bring the derivative inside f, because $f(A)$ is a linear expression with respect to the entries of A, and is without constant terms. The computations for $\dfrac{d}{d\theta}B_\theta^{(ik)}\Big|_{\theta=0}$, etc. are easy.

This Hessian is an $\left(\dfrac{m(m-1)}{2} \times \dfrac{m(m-1)}{2}\right)$ matrix, and is diagonal. The diagonal entries are non-zero. Therefore, we have found that A is a non-degenerate critical point.

The "(i,k)-th diagonal entry" is $-c_i\varepsilon_i - c_k\varepsilon_k$. We urge the reader to find the number of minus signs on the diagonal of this Hessian (Exercise 3.4). The answer is as follows:

Suppose that the subscripts i of the diagonal entries ε_i of A, $1 \leq i \leq m$, with $\varepsilon_i = 1$, are

$$i_1, i_2, \cdots, i_n$$

in ascending order. Then the index of the Hessian at A (the number of minus signs on the diagonal) is equal to

(3.57) $(i_1 - 1) + (i_2 - 1) + \cdots + (i_n - 1)$.

(The index is 0 if all the ε_i's are -1.) Also, the critical value at the critical point is

(3.58) $$2(c_{i_1} + c_{i_2} + \cdots + c_{i_n}) - \sum_{i=0}^{m} c_i.$$

It is easy to see that the critical values are distinct for different critical points if

(3.59) $2c_i < c_{i+1}, \quad i = 1, \cdots, m-1$.

Considering the condition $\det A = 1$, there are 2^{m-1} critical points (not 2^m).

As a special case, for $SO(3)$, the Morse function $f : SO(3) \to \mathbb{R}$ has 4 critical points. With the assumtion (3.59) on the c_i's, their indices are

$$0, \ 1, \ 2, \ 3$$

in the ascending order of the critical values.

REMARK. The manifold $SO(3)$ has the same number of critical points as P^3, and in fact, $SO(3)$ and P^3 are diffeomorphic. This can be proved from the fact that the special unitary group $SU(2)$ (see Example 3.14) of degree 2 is diffeomorphic to S^3 by Corollary 3.16, and the fact that the "adjoint representation $SU(2) \to SO(3)$" is a 2-fold covering.

For readers who know about Hamilton quaternions, $SU(2)$ is the set of quaternions q of norm 1, and $\mathbb{R}^3$ is the set of pure quaternions x. Then the adjoint representation is $SU(2) \to SO(3)$, $q \mapsto \{x \mapsto qx\bar{q}\}$, which is $2 : 1$ with kernel $\{\pm 1\}$.

EXAMPLE 3.13 (Unitary group $U(m)$). For a complex $m \times m$ matrix $U = (z_{ij})$, set $U^* = (\overline{z_{ji}})$. That is, U^* is obtained from U by taking the conjugate of all the entries, and taking the transpose at the same time. A complex $m \times m$ matrix U which satisfies

$$(3.60) \qquad\qquad U^* \, U = E$$

is called a *unitary matrix* of degree m. From the definition, the determinant of a unitary matrix U is a complex number with modulus 1: $|\det U| = 1$. The set $U(m)$ of all unitary matrices of degree m forms a Lie group under matrix product, and is called the *unitary group* of degree m.

The *hermitian product* between column vectors $\mathbf{a}_i$, $\mathbf{a}_j$ is defined by

$$(3.61) \qquad\qquad \mathbf{a}_i \cdot \overline{\mathbf{a}_j} = z_{1i}\overline{z_{1j}} + z_{2i}\overline{z_{2j}} + \cdots + z_{mi}\overline{z_{mj}},$$

and can be used to express the condition for a unitary matrix as $\mathbf{a}_i \cdot \overline{\mathbf{a}_i} = 1 \ (i = 1, \cdots, m)$, $\mathbf{a}_i \cdot \overline{\mathbf{a}_j} = 0 \ (i \neq j)$.

As the first column $\mathbf{a}_1$, we can choose an arbitrary vector in $\mathbb{C}^m$ with unit length. The dimension of $\mathbb{C}^m$ as a real vector space is $2m$, and the unit sphere in this space has dimension $2m - 1$, so that the degree of freedom for the choice of $\mathbf{a}_1$ is $(2m - 1)$-dimensional. Once $\mathbf{a}_1$ is chosen, an arbitrary unit vector can be selected as $\mathbf{a}_2$ from the complex $(m - 1)$-dimensional space that is orthogonal to $\mathbf{a}_1$ with respect to the hermitian product, and this freedom contributes

$\{2(m-1)-1\}$ dimensions. Repeating the same argument, we obtain that the dimension of $U(m)$ as a manifold is equal to

$$(3.62) \qquad \{2m-1\} + \{2(m-1)-1\} + \cdots + 3 + 1 = m^2.$$

Similarly to the case of $SO(m)$, for real numbers $1 < c_1 < c_2 < \cdots c_m$, define a function $f : U(m) \to \mathbb{R}$ by

$$(3.63) \qquad f(U) = \Re(c_1 z_{11} + c_2 z_{22} + \cdots + c_m z_{mm}),$$

where $\Re$ denotes the real part, and

$$(3.64) \qquad U = \begin{pmatrix} z_{11} & z_{12} & \cdots & z_{1m} \\ z_{21} & z_{22} & \cdots & z_{2m} \\ & \cdots & \cdots & \\ z_{m1} & z_{m2} & \cdots & z_{mm} \end{pmatrix}.$$

Although the following arguments are almost identical to those for $SO(m)$, we use the following matrices $C_\theta^{(ik)}$ ($1 \le i < k \le m$) and $A_\theta^{(i)}$ ($1 \le i \le m$), in addition to the rotation matrices $B_\theta^{(ik)}$ that we used for $SO(m)$. The matrix $C_\theta^{(ik)}$ is defined in terms of entries z_{pq} by

$$(3.65) \qquad z_{pq} = \begin{cases} \cos\theta & (p,q) = (i,i) \text{ or } (k,k), \\ \sqrt{-1}\,\sin\theta & (p,q) = (i,k) \text{ or } (k,i), \\ 1 & (p,q) = (j,j) \text{ where } j \ne i,k, \\ 0 & \text{otherwise.} \end{cases}$$

For example,

$$C_\theta^{(12)} = \begin{pmatrix} \cos\theta & \sqrt{-1}\,\sin\theta & \cdots & 0 \\ \sqrt{-1}\,\sin\theta & \cos\theta & & 0 \\ & & \ddots & \\ \vdots & \vdots & 1 & \\ & & & \ddots \\ 0 & 0 & \cdots & 1 \end{pmatrix}.$$

Also, the matrix $A_\theta^{(i)}$ is defined by

$$(3.66) \qquad A_\theta^{(i)} = \begin{pmatrix} 1 & 0 & & \cdots & & 0 \\ 0 & 1 & & \cdots & & 0 \\ & & \ddots & & & \\ & & & \exp\sqrt{-1}\,\theta & & \\ & & & & \ddots & \\ 0 & 0 & & \cdots & & 1 \end{pmatrix}.$$

(The diagonal entries are 1 except the i-th entry $\exp \sqrt{-1}\,\theta$, and all the other entries are 0's.)

There are m such matrices $A_\theta^{(i)}$. There are total of

$$\frac{m(m-1)}{2} + \frac{m(m-1)}{2} = m(m-1)$$

matrices of types $B_\theta^{(ik)}$ and $C_\theta^{(ik)}$, so in total there are $m + m(m-1)$ $= m^2$ matrices of the three types A, B, and C. This equals the dimension of $U(m)$. The tangent space of $U(m)$ is spanned by the rotation directions of these matrices A, B, C, and the same argument as in the case of $SO(m)$ can be applied.

The result is that the critical points of $f : U(m) \to \mathbb{R}$ are the unitary matrices with ± 1 on the diagonal entries, as in the statement of Lemma 3.12. However, unlike the case for rotation groups, it is not required that $\det = 1$.

The Hessian of $f : U(m) \to \mathbb{R}$ at a unitary matrix U, with $\varepsilon_1, \varepsilon_2, \cdots, \varepsilon_m$ $(\varepsilon_i = \pm 1)$ on the diagonal and 0 elsewhere, is diagonal and non-degenerate, and its index is computed as follows. Let

$$i_1, \ i_2, \ \cdots, \ i_n$$

be the subscripts i with $\varepsilon_i = 1$ in ascending order. Then the index is

$$(2i_1 - 1) + (2i_2 - 1) + \cdots + (2i_n - 1).$$

The index is 0 if all ε_i's are -1.

EXAMPLE 3.14 (Special unitary group $SU(m)$). The determinant of a unitary matrix is a complex number $\exp(\sqrt{-1}\,\theta)$ with modulus 1, but in particular, a unitary matrix with determinant 1 is called a special unitary matrix. The Lie group formed by all $m \times m$ special unitary matrices is called the *special unitary group* of degree m, and denoted by $SU(m)$. The condition $\det = 1$ reduces the dimension by 1 from the dimension of $U(m)$, so that

$\dim SU(m) = \dim U(m) - 1 = m^2 - 1$. For $m = 1$, $SU(1)$ consists of a single point (the trivial group). From now on, we assume that $m \geq 2$.

LEMMA 3.15. *Let* $f = \Re(c_1 z_{11} + c_2 z_{22} + \cdots + c_m z_{mm})$ *be the Morse function* $f : U(m) \to \mathbb{R}$ *constructed in Example 3.13 (where* $1 < c_1 < c_2 < \cdots < c_m$*). If the* c_i *(*$i = 2, \cdots, m$*) are sufficiently large comparing to* c_1*, then the restriction* $f|SU(m) : SU(m) \to \mathbb{R}$ *of* f *on* $SU(m)$ *is a Morse function on* $SU(m)$*. Its critical points are the diagonal matrices with* ± 1 *on the diagonal and with determinant* 1.

PROOF. Of the matrices $B_\theta^{(ik)}$, $C_\theta^{(ik)}$ and $A_\theta^{(i)}$ we used for the unitary matrices, $B_\theta^{(ik)}$ and $C_\theta^{(ik)}$ are special unitary matrices. Although the $A_\theta^{(i)}$ are not, if we set

$$D_\theta^{(i)} = A_\theta^{(i)} \cdot \{A_\theta^{(1)}\}^{-1}, \quad i = 2, \cdots, m,$$

then $D_\theta^{(i)}$ is a special unitary matrix. The number of the matrices $B_\theta^{(ik)}$, $C_\theta^{(ik)}$, $D_\theta^{(i)}$ in total is

$$\frac{m(m-1)}{2} + \frac{m(m-1)}{2} + (m-1) = m^2 - 1$$

and matches the dimension of $SU(m)$. Using the matrices $B_\theta^{(ik)}$ and $C_\theta^{(ik)}$ and a similar argument to that for Lemma 3.12, we see that a critical point of $f|SU(m) : SU(m) \to \mathbb{R}$ is a diagonal matrix A. Furthermore, as this diagonal matrix A is a special unitary matrix, its diagonal entry z_{ii} is a complex number $\exp(\sqrt{-1}\,\theta_i)$ with modulus 1, so that their product is 1:

$$\text{(3.67)} \quad \begin{aligned} &\exp(\sqrt{-1}\,\theta_1)\exp(\sqrt{-1}\,\theta_2)\cdots\exp(\sqrt{-1}\,\theta_m) \\ &= \exp(\sqrt{-1}\,(\theta_1 + \theta_2 + \cdots + \theta_m)) = 1. \end{aligned}$$

There is one more condition for this matrix A to be a critical point of $f|SU(m)$, which is

$$\begin{aligned} &\frac{d}{d\theta} f(A \cdot D_\theta^{(i)}) \bigg|_{\theta=0} \\ &= \Re(-c_1\sqrt{-1}\,\exp(\sqrt{-1}\,\theta_1) + c_i\sqrt{-1}\,\exp(\sqrt{-1}\,\theta_i)) \\ &= 0, \quad i = 2, \cdots, m. \end{aligned}$$

This condition is equivalent to

$$\text{(3.68)} \quad c_1 \sin\theta_1 = c_i \sin\theta_i, \quad i = 2, \cdots, m.$$

Under the condition that the c_i $(i = 2, \cdots, m)$ are sufficiently large compared to c_1, the above two conditions (3.67) and (3.68) can be combined to derive the condition

$$(3.69) \qquad \sin\theta_1 = \sin\theta_2 = \cdots = \sin\theta_m = 0.$$

We prove this. If $\sin\theta_1 \neq 0$, then by (3.68), we obtain $\sin\theta_i \neq 0$. However, the absolute value of $\sin\theta_i$ is sufficiently small compared to that of $\sin\theta_1$, if c_i is sufficiently large compared to c_1. (For example, if we assume that $100mc_1 < c_i$, say, then we have $|\sin\theta_i| < \dfrac{1}{100m}|\sin\theta_1|$.) In the following arguments, we only need the absolute values of sin, and the absolute values $|\sin\theta_i|$ do not change by adding to the angle θ_i an integral multiple of π, so we may assume that

$$-\frac{\pi}{2} \leq \theta_i \leq \frac{\pi}{2}.$$

Then from $|\sin\theta_i| < \dfrac{1}{100m}|\sin\theta_1|$ we obtain

$$|\theta_i| < \frac{1}{10m}|\theta_1|, \quad i = 2, \cdots, m,$$

even by a rough estimate. With these estimates, however, we never get $\sin(\theta_1 + \theta_2 + \cdots + \theta_m) = 0$. Thus a contradiction to (3.67) is obtained, and this implies the equalities (3.69).

We already stated that a critical point A of the function $f|SU(m)$ is a diagonal matrix with $\exp\sqrt{-1}\,\theta_i$ on the diagonals, but according to (3.69), the diagonal entries are in fact ± 1. Let these be $\varepsilon_1, \varepsilon_2, \cdots, \varepsilon_m$ $(\varepsilon_i = \pm 1)$.

By mutiplying the matrix A from the right by all $B_\theta^{(ik)}, \cdots,$ $C_\varphi^{(jl)}, \cdots, D_\psi^{(h)}, \cdots,$ the matrices

$$(3.70) \qquad A \cdot B_\theta^{(ik)} \cdots C_\varphi^{(jl)} \cdots D_\psi^{(h)} \cdots$$

form a local coordinate system $(\theta, \cdots, \varphi, \cdots, \psi, \cdots)$ of $SU(m)$ about A. Unlike the case for the unitary groups, the Hessian $H_{f|SU(m)}(A)$ of $f|SU(m)$ at A with respect to this local coordinate system contains non-diagonal $(m-1) \times (m-1)$ matrices, corresponding to the directions of the tangent vectors of $\{D_\theta^{(i)}\}_{i=2,\cdots,m}$.

By computing the second order derivatives

$$\left\{ \frac{\partial^2}{\partial\theta\partial\varphi} f(A \cdot D_\theta^{(i)} D_\varphi^{(j)}) \,\bigg|_{(\theta,\varphi)=(0,0)} \right\}_{i,j=2,\cdots,m},$$

we find that the non-diagonal portion of the matrix has the form

$$(3.71) \quad \begin{pmatrix} -c_2\varepsilon_2 - c_1\varepsilon_1 & -c_1\varepsilon_1 & \cdots & -c_1\varepsilon_1 \\ -c_1\varepsilon_1 & -c_3\varepsilon_3 - c_1\varepsilon_1 & \cdots & -c_1\varepsilon_1 \\ & & \ddots & \\ -c_1\varepsilon_1 & -c_1\varepsilon_1 & \cdots & -c_m\varepsilon_m - c_1\varepsilon_1 \end{pmatrix}.$$

Assuming that the c_i $(i = 2, \cdots, m)$ are sufficiently large compared to c_1 (say, assuming $c_i > m!\ c_1$), the non-diagonal entries of the matrix (3.71) are sufficiently small compared to the diagonal entries, so that the properties of the matrix resemble those of the diagonal matrix with the diagonal entries

$$-c_2\varepsilon_2, \ -c_3\varepsilon_3, \ \cdots, \ -c_m\varepsilon_m.$$

In particular, the determinant is non-zero, and the number of minus signs of the diagonalized matrix is equal to the number of minus signs among $-c_2\varepsilon_2, -c_3\varepsilon_3, \cdots, -c_m\varepsilon_m$.

In the Hessian $H_{f|SU(m)}(A)$, the portion other than the above portion (3.71) (the portion containing the tangent vectors of $B_\theta^{(ik)}$ and $C_\varphi^{(jl)}$) are diagonal matrices with the same entries as those of $f : U(m) \to \mathbb{R}$, so it follows that $\det H_{f|SU(m)}(A) \neq 0$. Therefore, A is a non-degenerate critical point, and $f|SU(m) : SU(m) \to \mathbb{R}$ is a Morse function. This completes the proof of Lemma 3.15. $\square$

We see from Lemma 3.15 that exactly half of the critical points of $f : U(m) \to \mathbb{R}$ are the critical points of $f|SU(m) : SU(m) \to \mathbb{R}$. That is, the diagonal matrices A with $\varepsilon_1, \varepsilon_2, \cdots, \varepsilon_m$ $(\varepsilon_i = \pm 1)$, and with $\det A = \varepsilon_1 \cdots \varepsilon_m = 1$, are the critical points of $f|SU(m) : SU(m) \to \mathbb{R}$. From the fact stated at the end of the proof of Lemma 3.15, the diagonalized matrix $\mathcal{H}_{f|SU(m)}(A)$ of the Hessian $H_{f|SU(m)}(A)$ is essentially the same as the matrix obtained from the Hessian $H_f(A)$ of $f : U(m) \to \mathbb{R}$ (which is already diagonal) by deleting the diagonal entry $-c_1\varepsilon_1$. Therefore, the number of minus signs on the diagonal coincide for $\mathcal{H}_{f|SU(m)}(A)$ and $H_f(A)$ if $-c_1\varepsilon_1 > 0$, and the number for $\mathcal{H}_{f|SU(m)}(A)$ is one less than that for $H_f(A)$ if $-c_1\varepsilon_1 < 0$, since $\mathcal{H}_{f|SU(m)}(A)$ does not have this diagonal entry.

Combining this fact and the results for $U(m)$, we can determine the indices of the critical points for $SU(m)$. That is, if

$$i_1, i_2, \cdots, i_k$$

are the subscripts i with $\varepsilon_i = 1$ in ascending order, the index of the Morse function $f|SU(m)$ at the critical point A is given as follows.

If $\varepsilon_1 = -1$, then the index is

$$(3.72) \qquad \{2i_1 - 1\} + \{2i_2 - 1\} + \cdots + \{2i_k - 1\},$$

and if $\varepsilon_1 = 1$, then the index is

$$(3.73) \qquad \{2i_1 - 1\} + \{2i_2 - 1\} + \cdots + \{2i_k - 1\} - 1.$$

REMARK. If $\varepsilon_1 = 1$ then $i_1 = 1$, so that the expression (3.73) is equal to

$$(3.74) \qquad \{2i_2 - 1\} + \cdots + \{2i_k - 1\}.$$

COROLLARY 3.16. $SU(2)$ *is diffeomorphic to the* 3-*dimensional sphere* S^3.

PROOF. $SU(2)$ is a compact 3-dimensional manifold. By Lemma 3.15, $f|SU(2)$ has only two critical points

$$(3.75) \qquad \begin{pmatrix} 1 & 0 \\ 0 & 1 \end{pmatrix}, \quad \begin{pmatrix} -1 & 0 \\ 0 & -1 \end{pmatrix}.$$

Their indices are 3 and 0, respectively. Hence by Theorem 3.6, we obtain $SU(2) \cong S^3$. □

COROLLARY 3.17. $SU(m)$ *is simply connected. (See Section 5.1 for the definition of being "simply connected.")*

PROOF. For the critical points of the Morse function $f|SU(m)$, the smallest index is of course 0, and the critical point with the second smallest index is

$$\begin{pmatrix} -1 & 0 & \cdots & & & 0 \\ 0 & 1 & \cdots & & & 0 \\ & & -1 & & & \\ & & & \ddots & & \\ 0 & 0 & \cdots & & & -1 \end{pmatrix},$$

if m is odd, and

$$\begin{pmatrix} 1 & 0 & \cdots & & & 0 \\ 0 & 1 & \cdots & & & 0 \\ & & -1 & & & \\ & & & \ddots & & \\ 0 & 0 & \cdots & & & -1 \end{pmatrix},$$

if m is even. The index of the critical point is 3 in either case, so the handle decomposition associated with $f|SU(m)$ does not have 1-handles. (In fact, it does not have 2-handles either.) By Corollary 5.9 in Chapter 5, $SU(m)$ is simply connected. $\square$

We prove the following proposition on the relations between the two groups $U(m)$ and $SU(m)$.

PROPOSITION 3.18. *(i) $U(m)$ is diffeomorphic to the direct product $SU(m) \times S^1$ as a smooth manifold.*

(ii) For $m \geq 2$, the structure of $U(m)$ as a Lie group is not isomorphic to the direct product of two groups $SU(m) \times S^1$.

PROOF. A map $h : U(m) \to SU(m) \times S^1$ of class C^∞ is constructed as follows. For any unitary matrix U of degree m, define

$$h(U) = (U \cdot \{A_\theta^{(1)}\}^{-1}, \det U).$$

If U is a unitary matrix, $\det U$ is a complex number $\exp(\sqrt{-1}\,\theta)$ of modulus 1. The θ in the matrix $A_\theta^{(1)}$ is this θ inside exp. In the above definition, the set of complex numbers with modulus 1 is identified with the circle S^1. Also the matrix $A_\theta^{(1)}$ is the $m \times m$ diagonal matrix $A_\theta^{(i)}$ with $i = 1$ that appeared in Example 3.13 (the unitary group $U(m)$).

The inverse map $k : SU(m) \times S^1 \to U(m)$, also of class C^∞, is defined by

$$k(U', \exp(\sqrt{-1}\,\theta)) = U' \cdot A_\theta^{(1)}$$

for any $m \times m$ unitary matrix U' and any $\exp(\sqrt{-1}\,\theta)$. It is easily seen that h and k are inverse mappings of each other, so that h is a diffeomorphism. This proves (i).

To prove (ii), we need the following lemma.

LEMMA 3.19. *The $m \times m$ diagonal matrix $\Delta(\zeta)$, with the same complex number ζ on the diagonal, commutes with all special unitary matrices of size m. Conversely, the matrices that commute with all special unitary matrices of size m are the diagonal matrices $\Delta(\zeta)$.*

The first half of this lemma is obvious, so we prove the second half. For simplicity we only consider the case $m = 2$. Assuming a matrix $X = \begin{pmatrix} a & b \\ c & d \end{pmatrix}$ commutes with specific special unitary matrices of size 2, $U_1 = \begin{pmatrix} 0 & -1 \\ 1 & 0 \end{pmatrix}$ and $U_2 = \begin{pmatrix} 0 & \sqrt{-1} \\ \sqrt{-1} & 0 \end{pmatrix}$, if we write out entries for the equations $XU_1 = U_1 X$ and $XU_2 = U_2 X$, we obtain $a = d$ and $b = c = 0$. Therefore, $X = \Delta(a)$. A similar argument proves the general case for any m.

Now we prove (ii) of the proposition. In general, for a group G and its subgroup H, the set of all elements g of G such that g commutes with all the elements of H forms a subgroup of G. Such a subgroup is called the *centralizer* of H in G. In particular, the centralizer of G itself in G is called the *center* of G. From the above lemma, we see the structure of the centralizer of $SU(m)$ in $U(m)$. That is, the centralizer is a subgroup of the form $\{\Delta(\exp(\sqrt{-1}\,\theta))\}$, and is homeomorphic to S^1. The intersection between this group and $SU(m)$ is the center of $SU(m)$, and is the cyclic group of order m written as $\{\Delta(\exp(2\pi\sqrt{-1}\,k/m))\}_{k=0,1,\cdots,m-1}$. Denote this group by C_m.

On the other hand, the structure of the group product $SU(m)\times S^1$ is given by

$$
(3.76) \qquad
\begin{aligned}
(U_1,\ \exp(\sqrt{-1}\,\theta_1))\cdot(U_2,\ \exp(\sqrt{-1}\,\theta_2)) \\
= (U_1 U_2,\ \exp(\sqrt{-1}\,(\theta_1 + \theta_2))).
\end{aligned}
$$

As we will show below, there is a unique subgroup $SU(m) \times \{1\}$ of $SU(m)\times S^1$ which is isomorphic to $SU(m)$, so we identify these. Then we see that the centralizer of $SU(m)$ in $SU(m) \times S^1$ is $C_m \times S^1$. This is homeomorphic to the disjoint union of m circles.

Since the structures of the centralizers are different, $U(m)$ and $SU(m) \times S^1$ are not isomorphic as Lie groups.

Finally, we prove that if H is a subgroup of $SU(m) \times S^1$ isomorphic to $SU(m)$, then

$$
H = SU(m) \times \{1\}.
$$

It is sufficient to prove that the image of H under the projection $p : SU(m) \times S^1 \to S^1$ is $\{1\}$. By Corollary 3.17, H is simply connected. Therefore, if $i : H \to SU(m) \times S^1$ is the inclusion map, then the composition $p \circ i : H \to S^1$ lifts to a homomorphism $j : H \to \mathbb{R}$ (see [10]). That is, if $e : \mathbb{R} \to S^1$ is the map defined by $e(\theta) = \exp(\sqrt{-1}\,\theta)$, then $p \circ i$ can be written as $e \circ j$. Since the only compact subgroup of $\mathbb{R}$ is $\{1\}$, we obtain $j(H) = \{1\}$, and therefore, $p(H) = \{1\}$ is proved. □

As we have seen, handle decompositions of various manifolds can be obtained using Morse functions. The number and the indices of the critical points of a Morse function can be obtained fairly easily, but in general, it is difficult to obtain attaching maps of handles.

3.3. Sliding handles

The goal of this section is to explain the technique called "handle sliding." Using this technique, we can alter the attaching maps of handles without changing the diffeomorphism type of a handlebody.

Let M be an m-dimensional closed manifold, and $f : M \to \mathbb{R}$ a given Morse function. Using Theorem 2.34, we deform f if necessary so that different critical points have distinct critical values. Suppose there are $n + 1$ critical points

$$p_0, \, p_1, \, \cdots, \, p_n$$

arranged in ascending order of their critical values. Pick and fix a gradient-like vector field X for f.

By Theorem 3.4, M can be decomposed as a handlebody

$$(3.77) \qquad \begin{aligned} M = D^m \cup_{\varphi_1} D^{\lambda_1} &\times D^{m-\lambda_1} \cup_{\varphi_2} \cdots \\ &\cup_{\varphi_i} D^{\lambda_i} \times D^{m-\lambda_i} \cup_{\varphi_{i+1}} \cdots \cup_{\varphi_n} D^m, \end{aligned}$$

where λ_i is the index of the critical point p_i.

We look closely at the i-th handle $D^{\lambda_i} \times D^{m-\lambda_i}$ in the above handle decomposition. To simplify the notation we denote by

$$N_j$$

the "subhandlebody" $\mathcal{H}(D^m; \varphi_1, \cdots, \varphi_j)$ obtained by attaching handles from the 0-th handle through the j-th handle. Then the i-th handle is attached to N_{i-1} by an attaching map

$$(3.78) \qquad \varphi_i : \partial D^{\lambda_i} \times D^{m-\lambda_i} \to \partial N_{i-1}.$$

Then "sliding the handle $D^{\lambda_i} \times D^{m-\lambda_i}$" is to perturb the attaching map φ_i by an "isotopy" of ∂N_{i-1}. First we define isotopy.

DEFINITION 3.20 (Isotopy). Let K be a k-dimensional manifold. The set $\{h_t\}_{t \in J}$ is called a family of diffeomorphisms if to each real number t in an open interval J, a diffeomorphism $h_t : K \to K$ is assigned. The family $\{h_t\}_{t \in J}$ is called an *isotopy* of K if the following two conditions are satisfied.

(i) The open interval J contains the closed interval $[0, 1]$, and h_t is constantly the identity map on K, $h_t = h_0 = \mathrm{id}_K$, when the parameter t satisfies $t \leq 0$. Also, for any $t \geq 1$, h_t is always equal to h_1, $h_t = h_1 = h$, where h is a diffeomorphism of K.

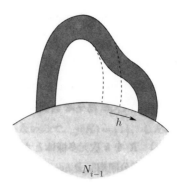

FIGURE 3.8. Sliding a handle

(ii) The map $H : K \times J \to K \times J$, defined by $H(x,t) = (h_t(x), t)$, is a diffeomorphism. The map h_t depends on the parameter t smoothly, in this sense.

We go back to the situation where a Morse function $f : M \to \mathbb{R}$ and a gradient-like vector field X for f are given. Let (3.77) be the handle decomposition determined by f and X.

THEOREM 3.21 (Sliding handles). *Fix one of the subscripts i of the critical points $(0 \le i \le n)$. Given an isotopy $\{h_t\}_{t \in J}$ of the boundary ∂N_{i-1} of a subhandlebody N_{i-1}, the attaching map φ_i of the handle $D^{\lambda_i} \times D^{m-\lambda_i}$ on N_{i-1} can be replaced by $h \circ \varphi_i$. Here h is the diffeomorphism h_1 corresponding to $t = 1$ of the isotopy $\{h_t\}_{t \in J}$. Also, by this replacement of the i-th attaching map, the diffeomorphism type of each of the subhandlebodies N_j $(0 \le j \le n)$ does not change.*

Figure 3.8 illustrates sliding a handle.

Theorem 3.21 can be explained more precisely as follows:

Recall that the handle decomposition (3.77) is determined by a Morse function f and a gradient-like vector field X for f. Then the theorem states that, given an isotopy $\{h_t\}_{t \in J}$ of ∂N_{i-1}, without changing the Morse function f, the gradient-like vector field X can be perturbed to a different gradient-like vector field Y, in such a way that the new handle decomposition determined by f and Y,

$$(3.79) \qquad \begin{aligned} M = D^m \cup_{\psi_1} D^{\lambda_1} \times D^{m-\lambda_1} \cup_{\psi_2} \cdots \\ \cup_{\psi_i} D^{\lambda_i} \times D^{m-\lambda_i} \cup_{\psi_{i+1}} \cdots \cup_{\psi_n} D^m, \end{aligned}$$

has the following properties (i), (ii), and (iii).

(i) The structure of the handle decomposition through the $(i-1)$-th handle, including the attaching maps, is unchanged. That is, in the new handle decomposition (3.79), if we denote by N'_j the subhandlebody from the 0-th handle up to the j-th handle, then $N'_j = N_j$ if j is less than or equal to $i-1$, and furthermore,

$$(3.80) \qquad \psi_j = \varphi_j.$$

(ii) The i-th attaching map is replaced by the following ψ_i:

$$(3.81) \qquad \psi_i = h \circ \varphi_i,$$

where $h : \partial N_{i-1} \to \partial N_{i-1}$ is the diffeomorphism corresponding to $t = 1$ in the given isotopy $\{h_t\}_{t \in J}$.

(iii) For any j $(0 \le j \le n)$, the diffeomorphism type of N'_j is the same as that of N_j:

$$(3.82) \qquad N'_j \cong N_j.$$

Now we prove Theorem 3.21.

PROOF. Let c_j be the value of f at the critical point p_j. We start with the handlebody (3.77). From the proof of Theorem 3.4, for each j $(0 \le j \le n)$, $M_{c_j+\varepsilon}$ can be identified with the subhandlebody $N_j = \mathcal{H}(D^m; \varphi_1, \cdots, \varphi_j)$, where ε is a sufficiently small positive number. We look closely at the i-th handle.

The attaching map (3.19) naturally determined where the i-th handle is attached to $M_{c_i-\varepsilon}$ is denoted by

$$(3.83) \qquad \varphi : \partial D^{\lambda_i} \times D^{m-\lambda_i} \to \partial M_{c_i-\varepsilon}.$$

Also consider the diffeomorphism

$$(3.84) \qquad \Phi : M_{c_{i-1}+\varepsilon} \to M_{c_i-\varepsilon}$$

given in the proof of Theorem 3.4. By identifying N_{i-1} and $M_{c_{i-1}+\varepsilon}$, the attaching map φ_i where the i-th handle is attached to N_{i-1} is given (cf. Theorem 3.4) by

$$(3.85) \qquad \varphi_i = \Phi^{-1} \circ \varphi : \partial D^{\lambda_i} \times D^{m-\lambda_i} \to \partial N_{i-1}.$$

We look at the diffeomorphism Φ more closely. Since f does not have critical points in between the values $c_{i-1} + \varepsilon$ and $c_i - \varepsilon$, Theorem 2.31 in the preceding chapter implies that $f^{-1}([c_{i-1} + \varepsilon, c_i - \varepsilon])$ is diffeomorphic to $\partial M_{c_{i-1}+\varepsilon} \times [0,1]$:

$$(3.86) \qquad f^{-1}([c_{i-1} + \varepsilon, c_i - \varepsilon]) \cong \partial M_{c_{i-1}+\varepsilon} \times [0,1].$$

Furthermore, when we identify the two sides of (3.86) by this diffeomorphism, at every point p of $\partial M_{c_{i-1}+\varepsilon}$ the interval $\{p\} \times [0,1]$ in the right-hand side of (3.86) corresponds to the integral curve of the gradient-like vector field X passing through p in the left-hand side.

In fact, the diffeomorphism in (3.86) can be defined on a larger region. This is because intervals a bit larger than $[c_{i-1} + \varepsilon, c_i - \varepsilon]$, say $[c_{i-1} + \varepsilon/2, c_i - \varepsilon/2]$, do not contain critical points of f either. Thus for a sufficiently small positive number δ, the diffeomorphism (3.86) can be extended to a diffeomorphism

$$(3.87) \qquad f^{-1}([c_{i-1} + \varepsilon/2, c_i - \varepsilon/2]) \cong \partial M_{c_{i-1}+\varepsilon} \times [-\delta, 1 + \delta].$$

Under the identification of both sides, it is still the case that $\{p\} \times [-\delta, 1 + \delta]$ in the right-hand side corresponds to an integral curve of X in the left-hand side.

The diffeomorphism Φ in question can be interpreted in terms of the right-hand side as a diffeomorphism that stretches $\partial M_{c_{i-1}+\varepsilon} \times [-\delta, 0]$ to $\partial M_{c_{i-1}+\varepsilon} \times [-\delta, 1]$. That is, it can be written as

$$(3.88) \quad \Phi : (p, t) \mapsto \left(p, \frac{1+\delta}{\delta} t + 1\right), \quad \forall (p, t) \in \partial M_{c_{i-1}+\varepsilon} \times [-\delta, 0].$$

On the boundary $\partial M_{c_{i-1}+\varepsilon}$, Φ is the map assigning $(p, 1)$ to $(p, 0)$ (in terms of the right-hand side of (3.87)).

By identifying N_{i-1} and $M_{c_{i-1}+\varepsilon}$, we regard an isotopy $\{h_t\}_{t \in J}$ as given on $\partial M_{c_{i-1}+\varepsilon}$. Then by the definition of an isotopy, the map $H : \partial M_{c_{i-1}+\varepsilon} \times J \to \partial M_{c_{i-1}+\varepsilon} \times J$ defined by

$$H(x, t) = (h_t(x), t)$$

is a diffeomorphism. From the condition (i) of an isotopy, H is "constant" independent of t for $t \leq 0$ or $t \geq 1$. For later convenience we consider

$$\tilde{H}(x, t) = (h_{1-t}(x), t)$$

obtained from H by putting it upside down. The map $\tilde{H} : \partial M_{c_{i-1}+\varepsilon} \times J \to \partial M_{c_{i-1}+\varepsilon} \times J$ is also a diffeomorphism, and is constant on $t \leq 0$ and $t \geq 1$. Since J is an open interval containing $[0,1]$, by adjusting

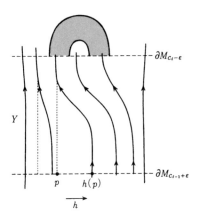

Y

p $h(p)$

$\xrightarrow{\quad}$
h

$\partial M_{c_i - \varepsilon}$

$\partial M_{c_{i-1} + \varepsilon}$

FIGURE 3.9. Integral curves of Y

the length, we can identify J with the open interval $(-\delta, 1 + \delta)$ in $[-\delta, 1 + \delta]$ which appears in the right-hand side of (3.87).

In (3.87) where the two sides are identified, there is a gradient-like vector field X on an open set $\partial M_{c_{i-1}+\varepsilon} \times (-\delta, 1 + \delta)$, and its integral curves are $\{p\} \times (-\delta, 1 + \delta)$. Hence X can be regarded as the vector field

$$\frac{\partial}{\partial t}$$

on this open subset. Consider the vector field $\tilde{H}_*(X)$, induced from X by the diffeomorphism $\tilde{H} : \partial M_{c_{i-1}+\varepsilon} \times (-\delta, 1 + \delta) \to \partial M_{c_{i-1}+\varepsilon} \times (-\delta, 1 + \delta)$. Since $\tilde{H}$ is constant on the ranges $t \leq 0$ and $t \geq 1$, $\tilde{H}_*(X)$ is $\frac{\partial}{\partial t}(= X)$ on this range. Thus, if we replace the vector field X by $\tilde{H}_*(X)$ in the open set $\partial M_{c_{i-1}+\varepsilon} \times (-\delta, 1 + \delta)$, the new $\tilde{H}_*(X)$ extends smoothly to the original X outside of the open set $\partial M_{c_{i-1}+\varepsilon} \times (-\delta, 1 + \delta)$, since $\tilde{H}_*(X)$ coincides with X on the portion along which it is glued (the ranges $t \leq 0$ and $t \geq 1$). Let Y be the new vector field constructed by this perturbation. The integral curves of Y on the open set $\partial M_{c_{i-1}+\varepsilon} \times (-\delta, 1 + \delta)$ are as depicted in Figure 3.9.

In the same way as the diffeomorphism $\Phi : M_{c_{i-1}+\varepsilon} \to M_{c_i-\varepsilon}$ was determined by "flowing" $M_{c_{i-1}+\varepsilon}$ along the integral curves of the vector field X, a diffeomorphism

(3.89) $$\Psi : M_{c_{i-1}+\varepsilon} \to M_{c_i-\varepsilon}$$

is determined by flowing $M_{c_{i-1}+\varepsilon}$ along the integral curves of the vector field Y. By restricting Ψ on the boundary $\partial M_{c_{i-1}+\varepsilon}$, as we see from Figure 3.9, Ψ is the mapping which sends $(h(p), 0)$ to $(p, 1)$ (in the notation in the right-hand side of (3.87)), where p is any point of $\partial M_{c_{i-1}+\varepsilon}$. Therefore, in the handle decomposition determined by f and Y, the i-th handle is attached to $N_{i-1} = M_{c_{i-1}+\varepsilon}$ by the attaching map

(3.90) $$\Psi^{-1} \circ \varphi = h \circ \varphi_i.$$

The structure of the handlebodies up to the $(i-1)$-th handle remains unchanged, since X and Y coincide on that portion of the handlebodies. It is also obvious that the diffeomorphism type of $N_j = M_{c_j+\varepsilon}$ remains unchanged for any j, since the definition of $M_{c_j+\varepsilon}$ $(= \{\, p \in M \mid f(p) \le c_j + \varepsilon \,\})$ does not involve the gradient-like vector field. This completes the proof of Theorem 3.21 (sliding handles).   □

As an application of Theorem 3.21, we prove the following.

THEOREM 3.22 (Arrangements of critical points). *Let M be an m-dimensional closed manifold, and $f : M \to \mathbb{R}$ a Morse function on it. Then f can be perturbed in such a way that the following condition $(*)$ is satisfied after the perturbation.*

$(*)$ *For any critical points p_i and p_j, $f(p_i) < f(p_j)$ implies that $index\,(p_i) \le index\,(p_j)$, where $index\,(p)$ denotes the index of a critical point p.*

This theorem states that the critical points can be arranged in such a way that the indices increase (are non-decreasing, to be precise) as the critical values increase.

This theorem is about perturbations of functions, but the proof uses the handle decompositions of a given $f : M \to \mathbb{R}$, and is geometric. The proof will be given after we prove three lemmas.

LEMMA 3.23 (General position). *Let S_1 and S_2 be compact submanifolds of dimensions s_1 and s_2 respectively, in a k-dimensional manifold K. If*

(3.91) $$s_1 + s_2 < k,$$

then there is an isotopy $\{h_t\}_{t \in J}$ of K such that $h_0 = \mathrm{id}_K$ and

$$(3.92) \qquad\qquad h_1(S_1) \cap S_2 = \emptyset.$$

Namely, this lemma says that S_1 can be separated from S_2 by moving S_1 with an isotopy $\{h_t\}_{t \in J}$ of K if the condition (3.91) is satisfied. In other words, if the dimensions of S_1 and S_2 are small enough comparing to the dimension of K (so that the inequality (3.91) is satisfied), then the general mutual position for S_1 and S_2 is being disjoint from each other.

PROOF. For simplicity we assume that "S_1 has an open neighborhood U diffeomorphic to $S_1 \times \mathrm{int}(D^{k-s_1})$," where $\mathrm{int}(D^{k-s_1})$ denotes the interior of the $(k - s_1)$-dimensional disk D^{k-s_1}. Furthermore, assume that the submanifold S_1 corresponds to $S_1 \times \{0\}$ in the direct product $S_1 \times \mathrm{int}(D^{k-s_1})$, where 0 denotes the center of D^{k-s_1}. We may assume this condition in the above quotes " ", since in this book we use Lemma 3.23 only when this condition is satisfied.

Identifying U with the direct product $S_1 \times \mathrm{int}(D^{k-s_1})$, define the projection to the second factor of the product

$$(3.93) \qquad\qquad \pi : U \to \mathrm{int}(D^{k-s_1}).$$

By mapping $S_2 \cap U$ into $\mathrm{int}(D^{k-s_1})$ under this π, the assumption (3.91) implies that

$$(3.94) \qquad \dim(S_2 \cap U) = s_2 < k - s_1 = \dim(D^{k-s_1}),$$

so that the image $\pi(S_2 \cap U)$ is a subset of $\mathrm{int}(D^{k-s_1})$ of a low dimension, and therefore is a "nowhere dense subset." Namely, in any neighborhood of any point p of $\mathrm{int}(D^{k-s_1})$, there is a point which is not contained in $\pi(S_2 \cap U)$. (Strictly speaking, this fact follows from a special case of Sard's theorem in the case when there is a difference in dimensions, cf. [15].) Pick a point p_0 near the center 0 of D^{k-s_1} which is not contained in $\pi(S_2 \cap U)$.

There is an isotopy $\{j_t\}_{t \in J}$ of the interior of the disk $\mathrm{int}(D^{k-s_1})$ with the following properties (i) and (ii) (Excercise 3.2).

(i) For $t = 0$ and 1, $j_0 = \mathrm{id}$ and $j_1(0) = p_0$, respectively.

(ii) For any t, j_t is the identity map outside of $\frac{1}{2}D^{k-s_1}$, the concentric circle of D^{k-s_1} with half the radius.

Using this isotopy $\{j_t\}_{t\in J}$, we define an isotopy $\{h_t\}_{t\in J}$ on $U = S_1 \times \text{int}(D^{k-s_1})$ by

$$(3.95) \quad h_t(p, \mathbf{x}) = (p, j_t(\mathbf{x})), \quad \forall (p, \mathbf{x}) \in S_1 \times \text{int}(D^{k-s_1}), \quad \forall t \in J.$$

By the condition (ii) of the isotopy $\{j_t\}_{t\in J}$, the isotopy $\{h_t\}_{t\in J}$ constructed above is the identity mapping outside of the direct product $S_1 \times \frac{1}{2}D^{k-s_1}$, so that $\{h_t\}_{t\in J}$ can be extended to an isotopy on all of K by defining it to be the identity mapping outside of U. The extended isotopy is also denoted by $\{h_t\}_{t\in J}$. After moving S_1 by this isotopy, $h_1(S_1)$ is supposed to match $S_1 \times \{p_0\}$ in U, as we see from the construction (3.95). However, from the choice of p_0, $S_1 \times \{p_0\}$ does not intersect $S_2 \cap U$. Hence we have

$$h_1(S_1) \cap S_2 = \emptyset,$$

and this completes the proof of Lemma 3.23.   □

To continue the proof of the theorem, we give the following definition. Consider a situation where a Morse function $f : M \to \mathbb{R}$ and a gradient-like vector field X for f are given. Denote the critical value of a critical point p_i by c_i, and the index by λ_i.

DEFINITION 3.24 (Upper disk and lower disk). The set of all points p in

$$M_{[c_{i-1}+\varepsilon, c_i+\varepsilon]} = \{\, p \in M \mid c_{i-1} + \varepsilon \leq f(p) \leq c_i + \varepsilon \,\}$$

that converge to the critical point p_i along integral curves of X as the parameter t goes to infinity, $t \to +\infty$, is called the *lower disk* (or *left-hand disk*) *associated with the critical point* p_i. Here, the critical point p_i itself is considered to be included in the lower disk. Similarly, the set of all points in the same range $M_{[c_{i-1}+\varepsilon, c_i+\varepsilon]}$ that converge to p_i along an integral curve of X in the opposite direction as $t \to -\infty$ is called the *upper disk* (or *right-hand disk*) *associated with the critical point* p_i. Here, again, the critical point p_i itself is included in the upper disk (Figure 3.10).

The lower and upper disks associated with p_i are respectively denoted by

$$D_l(p_i), \quad D_u(p_i)$$

where subscripts l and u represent lower and upper, respectively.

REMARK. The names left-hand and right-hand disks may sound odd, but they are the terminology used in Milnor's book [14], and the names upper and lower disks are adopted for this book by the

author. In this book, the figures are depicted as flows going upwards
for gradient-like vector fields, so the terms upper and lower disks are
used instead.

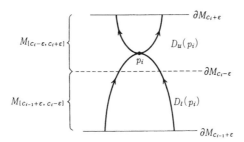

FIGURE 3.10. Upper and lower disks

From the fact that a gradient-like vector field X matches the gradient vector field in a neighborhood of p_i, and recalling the definition
of the lower disk $D_l(p_i)$ and the standard form (3.8) of f near the
critical point p_i, we see that

(3.96)
$$D_l(p_i) \cap M_{[c_i-\varepsilon,c_i+\varepsilon]}$$
$$= \{ (x_1,\cdots,x_m) \mid x_1^2 + \cdots + x_{\lambda_i}^2 \le \varepsilon,\ x_{\lambda_i+1} = \cdots = x_m = 0 \}.$$

This intersection is a λ_i-dimensional disk. Also we see that

(3.97)
$$D_l(p_i) \cap M_{[c_{i-1}+\varepsilon,c_i-\varepsilon]} =$$

the boundary of the above λ_i-dimensional disk $\times$ $[c_{i-1} + \varepsilon, c_i - \varepsilon]$.

Since the disk $D_l(p_i)$ is the union of (3.96) and (3.97), it is diffeomorphic to a λ_i-dimensional disk itself.

Similarly $D_u(p_i)$ is diffeomorphic to an $(m-\lambda_i)$-dimensional disk.

When we regard $M_{c_i+\varepsilon}$ as $M_{c_{i-1}+\varepsilon}$ with a λ_i-handle attached,
the lower disk $D_l(p_i)$ corresponds to the core of the λ_i-handle, and
the upper disk $D_u(p_i)$ corresponds to the co-core.

The boundary $\partial D_l(p_i)$ of the lower disk $D_l(p_i)$ is embedded in
$\partial M_{c_{i-1}+\varepsilon}$ by the attaching map φ_i of a λ_i-handle. Notice also that
the boundary $\partial D_u(p_{i-1})$ of the upper disk $D_u(p_{i-1})$ of the preceding
critical point p_{i-1} is an embedded sphere in the same $\partial M_{c_{i-1}+\varepsilon}$ as
well.

In this situation the following holds.

LEMMA 3.25 (Separation of handles). *Fix a subscript i of the critical points. If index $(p_{i-1}) \geq$ index (p_i), then the gradient-like vector field X can be perturbed to another gradient-like vector field Y, in such a way that by replacing the attaching map φ_i by ψ_i using this perturbation, we have*

$$(3.98) \qquad \psi_i(\partial D_l(p_i)) \cap \partial D_u(p_{i-1}) = \emptyset,$$

keeping the Morse function $f : M \to \mathbb{R}$ fixed. This perturbation does not alter the situation of handles from the 0-th through the $(i-1)$-th.

PROOF. Denote the indices of the critical points p_{i-1} and p_i by λ_{i-1} and λ_i respectively. By assumption, we have

$$(3.99) \qquad\qquad \lambda_{i-1} \geq \lambda_i.$$

Also, the dimensions of $D_u(p_{i-1})$ and $D_l(p_i)$ are

$$\dim(D_u(p_{i-1})) = m - \lambda_{i-1}, \quad \dim(D_l(p_i)) = \lambda_i.$$

Considering the dimensions of the boundary spheres of these disks, we see from (3.99) that

$$(3.100)$$
$$\dim(\partial D_u(p_{i-1})) + \dim(\partial D_l(p_i)) = (m - \lambda_{i-1} - 1) + (\lambda_i - 1)$$
$$= m - 2 + (\lambda_i - \lambda_{i-1}) < m - 1.$$

Since $\dim(\partial M_{c_{i-1}+\varepsilon}) = m - 1$, Lemma 3.23 (general position) implies that there exists an isotopy $\{h_t\}_{t \in J}$ of $\partial M_{c_{i-1}+\varepsilon}$ which separates the image $\varphi_i(\partial D_l(p_i))$ from $\partial D_u(p_{i-1})$. (The assumption made in the proof of Lemma 3.23, that "there exists a neighborhood of the form of a direct product," is satisfied for attaching spheres $\varphi_i(\partial D_l(p_i))$ of handles.) This isotopy satisfies the condition

$$(3.101) \qquad h_1(\varphi_i(\partial D_l(p_i))) \cap \partial D_u(p_{i-1}) = \emptyset.$$

By applying Theorem 3.21 (sliding handles) here, the attaching map φ_i of the i-th handle can be perturbed to $h_1 \circ \varphi_i$. If we denote the new attaching map by ψ_i, the condition (3.101) obviously implies that $\psi_i(\partial D_l(p_i)) \cap \partial D_u(p_{i-1}) = \emptyset$.

It is obvious from Theorem 3.21 that the situation remains unchanged for the handles up to (and including) the $(i-1)$-th handle. This completes the proof of Lemma 3.25. $\qquad\qquad\qquad\qquad \square$

We assume that a Morse function $f : M \to \mathbb{R}$ and a gradient-like vector field X for f are fixed in the following lemma as well.

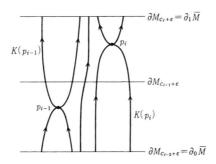

FIGURE 3.11. $K(p_{i-1})$ and $K(p_i)$

LEMMA 3.26 (Moving critical values up and down). *Consider the range*

$$(3.102) \qquad M_{[c_{i-2}+\varepsilon,\,c_i+\varepsilon]},$$

which contains two consecutive critical points p_{i-1}, p_i of a Morse function $f : M \to \mathbb{R}$. Let $K(p_{i-1})$ be the set of points p in (3.102) that converge to the critical point p_{i-1} along integral curves of X as the parameter t approaches infinity, $t \to \infty$ or $t \to -\infty$. We consider that p_{i-1} itself is contained in $K(p_{i-1})$. The set defined similarly for the critical point p_i is denoted by $K(p_i)$ (cf. Figure 3.11).

In this situation, if

$$(3.103) \qquad K(p_{i-1}) \cap K(p_i) = \emptyset,$$

then the Morse function $f : M \to \mathbb{R}$ can be perturbed to another Morse function $g : M \to \mathbb{R}$ such that g has the following properties:

(i) *g coincides with f outside of the region (3.102),*
(ii) *the sets of critical points and their indices coincide for f and g, and*
(iii) *$g(p_{i-1}) = a$ and $g(p_i) = b$,*

where a and b in condition (iii) are arbitrarily given real numbers that are larger than $c_{i-2} + \varepsilon$ and smaller than $c_i + \varepsilon$.

This lemma says that if the two sets $K(p_{i-1})$ and $K(p_i)$ do not intersect, then the functional values at the critical points can be moved up and down freely to match arbitrarily chosen real numbers a and b.

PROOF. To simplify the notation, set

(3.104)
$$\overline{M} = M_{[c_{i-2}+\varepsilon, c_i+\varepsilon]}, \quad \partial_0\overline{M} = f^{-1}(c_{i-2}+\varepsilon), \quad \partial_1\overline{M} = f^{-1}(c_i+\varepsilon).$$

The manifold $\overline{M}$ has boundary, which consists of the disjoint union of $\partial_0\overline{M}$ and $\partial_1\overline{M}$:

$$\partial\overline{M} = \partial_0\overline{M} \sqcup \partial_1\overline{M}.$$

The integral curve of X passing through a point p of $\overline{M} - K(p_{i-1}) \cup K(p_i)$ enters from $\partial_0\overline{M}$ and goes out through $\partial_1\overline{M}$ (cf. Figure 3.11).

Pick a function of class C^∞ on $\partial_0\overline{M}$,

(3.105) $$h : \partial_0\overline{M} \to \mathbb{R},$$

with the following two properties. Such a function is familiar in manifold theory (cf. [11] or [4], Chapter 1, §8, Exercise 15).

(A) $0 \le h \le 1$.

(B) h takes the value 0 on some open neighborhood of $K(p_{i-1}) \cap \partial_0\overline{M}$, and takes the value 1 on some open neighborhood of $K(p_i) \cap \partial_0\overline{M}$.

Using the function h, construct a function of class C^∞,

(3.106) $$\overline{h} : \overline{M} \to \mathbb{R},$$

on $\overline{M}$ as follows: for a point p in $\overline{M} - K(p_{i-1}) \cup K(p_i)$, the integral curve of X passing through p intersects $\partial_0\overline{M}$ at a unique point q, so set

$$\overline{h}(p) = h(q).$$

Define the value of $\overline{h}$ to be 0 on $K(p_{i-1})$, and 1 on $K(p_i)$. From the condition (B) of h, the function $\overline{h}$ is of class C^∞. Also, $\overline{h}$ takes a constant value on each integral curve of X (or on $K(p_{i-1})$ or on $K(p_i)$).

In what follows the phrase "integral curves" is used in a more general sense to save words. That is, when we say an "integral curve," we mean either an ordinary integral curve of X, or $K(p_{i-1})$, or $K(p_i)$.

The idea of the proof of Lemma 3.26 is first to construct a family $\{g_s\}_{s\in[0,1]}$ of strictly increasing functions of class C^∞ on the interval $[c_{i-2}+\varepsilon, c_i+\varepsilon]$ for the parameter s in the interval $[0,1]$, in such a way that the family depends on the parameter s ($0 \le s \le 1$) smoothly (of class C^∞). As we mentioned above, the value of $\overline{h}$ is constant on each integral curve. Looking closely at a single integral curve, if the constant value of $\overline{h}$ is s on the integral curve, then we define a new function g on the integral curve by $g(p) = g_s(f(p))$. The function

$g_s(f(p))$ increases strictly on the integral curve, since each $g_s(x)$ is strictly increasing with respect to x, and f is strictly increasing from $c_{i-2}+\varepsilon$ to $c_i+\varepsilon$ as a point moves along each integral curve from $\partial_0 \overline{M}$ to $\partial_1 \overline{M}$. If we define g_s in such a way that $g_0(f(p_{i-1})) = a$ for $s = 0$ and $g_1(f(p_i)) = b$ for $s = 1$, then the condition (iii) for g would also be satisfied. This is the idea of the proof.

Precisely speaking, the construction is as follows. Let

$$(3.107) \qquad G : [c_{i-2} + \varepsilon, c_i + \varepsilon] \times [0,1] \to [c_{i-2} + \varepsilon, c_i + \varepsilon]$$

be a function of class C^∞ with the following three conditions.

(a) $G(x, s)$ is strictly increasing as a function of x when a single value is fixed for s, and $G(x, s)$ increases from $c_{i-2} + \varepsilon$ to $c_i + \varepsilon$ as x increases from $c_{i-2} + \varepsilon$ to $c_i + \varepsilon$. (The function $g_s(x)$ mentioned in the idea of proof corresponds to $G(x, s)$.)

(b) $G(f(p_{i-1}), 0) = a$, $G(f(p_i), 1) = b$.

(c) For any s, $G(x, s) = x$ as long as x is in a sufficiently small neighborhood of $c_{i-2} + \varepsilon$ or $c_i + \varepsilon$. Furthermore, for x in a neighborhood of $f(p_{i-1})$,

$$\frac{\partial}{\partial x} G(x, 0) = 1,$$

and for x in a neighborhood of $f(p_i)$,

$$\frac{\partial}{\partial x} G(x, 1) = 1.$$

A proof of existence of such a function $G(x, s)$ is left to the reader as an exercise (cf. Exercise 3.3).

The new Morse function $g : M \to \mathbb{R}$ we seek is defined by

$$g(p) = G(f(p), \overline{h}(p)).$$

It is easily checked that g satisfies the conditions (i), (ii), and (iii).

This completes the proof of Lemma 3.26 (moving critical points up and down). $\square$

With the above preliminaries, we prove Theorem 3.22 for arranging critical points. Let $f : M \to \mathbb{R}$ be a given Morse function, and X a gradient-like vector field for f. Suppose that there are $n+1$ critical points and that they have distinct values. They are denoted by

$$p_0, \; p_1, \; \cdots, \; p_n,$$

in ascending order of critical values. The critical value of p_i is denoted by c_i, so that

$$c_0 < c_1 < \cdots < c_n.$$

Our proof would be done if the indices index(p_i) of p_i are already in ascending order, but suppose they are reversed, as

$$\text{index}(p_{i-1}) > \text{index}(p_i),$$

at a certain subscript i. By Lemma 3.25 (separation of handles), the vector field X associated with $f : M \to \mathbb{R}$ can be perturbed to Y so that the boundary $\partial D_l(p_i)$ of the lower disk $D_l(p_i)$ for p_i and the boundary $\partial D_u(p_{i-1})$ of the upper disk $D_u(p_{i-1})$ for p_{i-1} are disjoint in $f^{-1}(c_{i-1} + \varepsilon)$ $(= \partial M_{c_{i-1}+\varepsilon})$. Then $\partial D_l(p_i)$ does not converge to p_{i-1} even when it flows downward (into $M_{[c_{i-2}+\varepsilon, c_{i-1}+\varepsilon]}$) along Y, and $\partial D_u(p_{i-1})$ does not converge to p_i even when it flows upward (into $M_{[c_{i-1}+\varepsilon, c_i+\varepsilon]}$) along Y (cf. Figure 3.11).

Threfore, $K(p_{i-1})$ and $K(p_i)$ are disjoint in $M_{[c_{i-2}+\varepsilon, c_i+\varepsilon]}$, and Lemma 3.26 (moving critical points up and down) can be applied. Pick a and b in this lemma such that $a > b$; then $f : M \to \mathbb{R}$ can be perturbed in $M_{[c_{i-2}+\varepsilon, c_i+\varepsilon]}$ to $g : M \to \mathbb{R}$ such that g satisfies

$$g(p_{i-1}) = a > b = g(p_i).$$

The values of g remain unchanged from those of f for the critical points other than p_{i-1} and p_i. In summary, p_{i-1} used to have a larger index than p_i though it had a smaller functional value than p_i, but now, the functional value of p_{i-1} has been moved up to a higher position than p_i.

Thus p_{i-1} now has a larger index and a larger critical value than p_i, so we rename p_{i-1} and p_i by defining p_{i-1} to be p_i and p_i to be p_{i-1}, and set $a = c_i$, $b = c_{i-1}$. Then we get

$$g(p_{i-1}) < g(p_i), \quad \text{index}(p_{i-1}) < \text{index}(p_i),$$

and the reversal of critical values and indices between the $(i-1)$-th and the i-th critical points is corrected. By correcting the reversal, one step at a time, all reversals can be corrected after a finite number of steps, and the ascending order of the critical values and that of the indices are achieved.

This completes the proof of Theorem 3.22.

The situation after correcting reversals of indices is that, in terms of handle decompositions, a handlebody representing M is built by attaching handles, one at a time, of the same index as or a higher index than those of the previously attached handles.

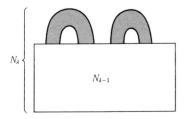

FIGURE 3.12. Handles of the same index can be attached simultaneously

In fact, if we look at the proof of Theorem 3.22 more closely, we see that the handles of the same index can be attached simultaneously, instead of one at a time. The next theorem states that.

THEOREM 3.27 (Simultaneous handle attachments). *For a given m-dimensional manifold M and a Morse function $f : M \to \mathbb{R}$, f can be modified to another Morse function $g : M \to \mathbb{R}$ such that g satisfies the following conditions (i), (ii), and (iii).*

(i) *The functions f and g have the same number of critical points. Furthermore, the number of critical points with the same index coincide for f and g as well.*

(ii) *For critical points p_i and p_j of g, $index(p_i) < index(p_j)$ implies $g(p_i) < g(p_j)$.*

(iii) *Also, index $(p_i) = $ index (p_j) implies $g(p_i) = g(p_j)$.*

Figure 3.12 illustrates the situation of the critical points of the Morse function g. As we see from this figure, N_λ is obtained from the handlebody $N_{\lambda-1}$, which consists of handles of indices less than or equal to $\lambda - 1$ all attached, by attaching a disjoint union of r copies of λ-handles

$$D^\lambda \times D^{m-\lambda} \sqcup \cdots \sqcup D^\lambda \times D^{m-\lambda}$$

at the same time. Here, r is the number of handles of index λ.

PROOF. We give an outline. In the proof of Theorem 3.22, we applied Lemma 3.25 (separation of handles) to the critical points p_{i-1} and p_i with $f(p_{i-1}) < f(p_i)$ and $index(p_{i-1}) > index(p_i)$, but Lemma 3.25 can be applied to p_{i-1} and p_i with $f(p_{i-1}) < f(p_i)$ and $index(p_{i-1}) = index(p_i)$, as well. Thus, when the indices of p_{i-1} and p_i coincide, we apply Lemma 3.25 and make $K(p_{i-1})$ and $K(p_i)$ disjoint. Then by Lemma 3.26 (moving critical values up and down), we can freely adjust the critical values of p_{i-1} and p_i, so that we perturb the function in such a way that the critical values of both of the critical points match. $\square$

3.4. Canceling handles

In this section, we describe a phenomenon where two consecutive critical points vanish under certain circumstances. As before, suppose that a closed m-dimensional manifold M and a Morse function f : $M \to \mathbb{R}$ are given, and a gradient-like vector field X for f is fixed. Suppose further that f has critical points

$$(3.108) \qquad\qquad p_0, \ p_1, \ \cdots, \ p_n,$$

their critical values are all distinct, and (3.108) is arranged in ascending order of the critical values. In other words, if we set $c_i = f(p_i)$, then

$$c_0 < c_1 < \cdots < c_n.$$

Fix a subscript i and consider the range $M_{[c_{i-2}+\varepsilon, c_i+\varepsilon]}$, which contains a pair of consecutive critical points. As in the proof of Lemma 3.26, we denote this range by $\overline{M}$.

THEOREM 3.28 (Canceling handles). *For the critical points p_{i-1} and p_i in $\overline{M}$, we make the following two assumptions.*

(i) *The index of p_i is one larger than the index of p_{i-1}. In other words, if we set $index(p_{i-1}) = \lambda$, then $index(p_i) = \lambda + 1$.*

(ii) *The boundary $\partial D_l(p_i)$ of the lower disk associated with p_i and the boundary $\partial D_u(p_{i-1})$ of the upper disk associated with p_{i-1} intersect transversely at a single point in the level surface $f^{-1}(c_{i-1} + \varepsilon)$, which separates the two critical points. (The meaning of "intersect transversely" will be explained below, but roughly it can be considered "intersect orthogonally." See p. 77.)*

Under the assumptions (i) and (ii), f can be perturbed to another Morse function g so that g satisfies the following two properties.

(A) *g does not have critical points in the interior of $\overline{M}$.*

(B) *g coincides with f near the boundary and the outside of $\overline{M}$.*

From (A), we see that the critical points p_{i-1} and p_i in $\overline{M}$ vanish together when f is perturbed to g.

Fix a gradient-like vector field Y for g. By (B), it can be assumed that Y coincides with X, a gradient-like vector field for f, outside of $\overline{M}$. In the handle decomposition determined by g and Y, the handle which used to correspond to p_{i-1} and the handle which used to correspond to p_i vanish together.

Before we start the proof, we define "intersecting transversely."

DEFINITION 3.29. Suppose that there are an a-dimensional manifold A and a b-dimensional manifold B in a k-dimensional manifold K such that $k = a + b$. We say that A and B *intersect transversely* at a point q_0 of K if there exist an open neighborhood U and a local coordinate system $(x_1, \cdots, x_k)$ of U such that $A \cap U$ is written as $x_{a+1} = x_{a+2} = \cdots = x_k = 0$, and $B \cap U$ can be written as $x_1 = x_2 = \cdots = x_a = 0$ in these local coordinates. (From this condition, we see that q_0 is the origin $(0, \cdots, 0)$ of the local coordinate system $(x_1, \cdots, x_k)$.)

The level surface $f^{-1}(c_{i-1} + \varepsilon)$ which appears in Theorem 3.28 is an $(m-1)$-dimensional manifold, and the manifolds $\partial D_l(p_i)$ and $\partial D_u(p_{i-1})$ embedded in it have dimensions λ and $m - \lambda$ respectively. Therefore

$$\dim f^{-1}(c_{i-1} + \varepsilon) = \dim \partial D_l(p_i) + \dim \partial D_u(p_{i-1}).$$

Thus, in terms of these dimensions, the boundary of the lower disk $D_l(p_i)$ and the boundary of the upper disk $D_u(p_{i-1})$ certainly could intersect tranversely at a single point in the level surface $f^{-1}(c_{i-1}+\varepsilon)$.

We outline a proof of Theorem 3.28.

PROOF. By assumption, $\partial D_l(p_i)$ and $\partial D_u(p_{i-1})$ intersect transversely at a single point in the level surface $f^{-1}(c_{i-1} + \varepsilon)$. Let q_0 be the intersection point; then the integral curve $C = C(t)$ of X passing through q_0 has the following property. Namely, points of C converge to the critical point p_i ($C(t) \to p_i$) as $t \to +\infty$, and converge to p_{i-1} ($C(t) \to p_{i-1}$) as $t \to -\infty$. Also, the only integral curve of X with this property is C. In what follows, C is considered to contain its end points p_{i-1} and p_i, and is considered as a compact curve segment.

Since the index of the critical point p_{i-1} is λ, from the Definition 2.29 of gradient-like vector fields in Chapter 2, X can be written as

(3.109)
$$
\begin{aligned}
X = {}& -2x_1 \frac{\partial}{\partial x_1} - \cdots - 2x_\lambda \frac{\partial}{\partial x_\lambda} \\
& + 2x_{\lambda+1} \frac{\partial}{\partial x_{\lambda+1}} + \cdots + 2x_m \frac{\partial}{\partial x_m}
\end{aligned}
$$

in a neighborhood V_1 of p_{i-1} with an appropriate local coordinate system $(x_1, \cdots, x_m)$. For later convenience we interchange the roles of x_1 and $x_{\lambda+1}$ and assume that X has the form of

(3.110)
$$X = 2x_1 \frac{\partial}{\partial x_1} - \cdots - 2x_\lambda \frac{\partial}{\partial x_\lambda}$$
$$- 2x_{\lambda+1} \frac{\partial}{\partial x_{\lambda+1}} + \cdots + 2x_m \frac{\partial}{\partial x_m}$$

near p_{i-1}.

Since the index of the critical point p_i is $\lambda + 1$, it can be written as

(3.111)
$$X = -2y_1 \frac{\partial}{\partial y_1} - \cdots - 2y_{\lambda+1} \frac{\partial}{\partial y_{\lambda+1}}$$
$$+ 2y_{\lambda+2} \frac{\partial}{\partial y_{\lambda+2}} + \cdots + 2y_m \frac{\partial}{\partial y_m}$$

in a neighborhood V_2 of p_i with an appropriate local coordinate system $(y_1, \cdots, y_m)$.

The descriptions (3.110) and (3.111) are those at neighborhoods V_1 and V_2 near the end points of the integral curve C, but in fact, we can find a local coordinate system $(x_1, \cdots, x_m)$ in an open neighborhood U of the entire curve C such that the following (i), (ii), and (iii) hold.

(i) The coordinates of p_{i-1} in this local coordinate system are $(0, 0, \cdots, 0)$.

(ii) The coordinates of p_i are $(1, 0, \cdots, 0)$.

(iii) X can be written on U as

(3.112)
$$X = 2v(x_1) \frac{\partial}{\partial x_1} - 2x_2 \frac{\partial}{\partial x_2} - \cdots - 2x_{\lambda+1} \frac{\partial}{\partial x_{\lambda+1}}$$
$$+ 2x_{\lambda+2} \frac{\partial}{\partial x_{\lambda+2}} + \cdots + 2x_m \frac{\partial}{\partial x_m}.$$

Here $v(x_1)$ is a one-variable function of class C^∞ with the variable x_1 defined on the interval $-\delta < x_1 < 1 + \delta$, where δ is a small positive number. Furthermore, $v(x_1) = x_1$ if x_1 is in a neighborhood of 0, and $v(x_1) = 1 - x_1$ if x_1 is in a neighborhood of 1. Also, $v(x_1) > 0$ in $0 < x_1 < 1$ (cf. Figure 3.13).

From the conditions on $v(x_1)$, the description (3.112) of X coincides with (3.110) in a neighborhood of p_{i-1}, and coincides with (3.111) in a neighborhood of p_i (after the coordinate change

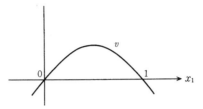

FIGURE 3.13. A function v

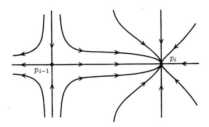

FIGURE 3.14. A vector field X

$(y_1, y_2, \cdots, y_m) = (x_1 - 1, x_2, \cdots, x_m))$. Figure 3.14 illustrates the situation of X in U.

The hardest part of the proof of Theorem 3.28 is to construct a local coordinate system $(x_1, \cdots, x_m)$ with the above properties in a neighborhood U of C. A careful argument is needed to patch together the neighborhoods V_1 and V_2 near both ends of C by perturbing them along the integral curve of X. In Milnor's book [14] a detailed proof is given in 12 pages, but for now we skip the proof and go ahead, taking this fact for granted.

We perturb the vector field X described in (3.112) in the neighborhood U of C so that $X \neq \mathbf{0}$ everywhere in U. For this purpose, we construct a family of functions $\{v_s(x_1)\}_s$ that depend on the parameter s smoothly. More precisely, the family is to satisfy the following conditions.

(i) Each $v_s(x_1)$ is a function of class C^∞ defined in $-\delta < x_1 < 1 + \delta$.

(ii) For a sufficiently small positive number η, the $v_s(x_1)$ are defined as long as the parameter s moves in $-\eta < s < 2\eta$.

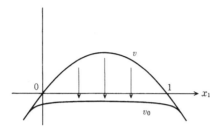

FIGURE 3.15. A perturbation of the function $v(x_1)$

(iii) For any $s \geq \eta$, we have $v_s(x_1) = v(x_1)$, where $v(x_1)$ is the function in the description (3.112) of X.

(iv) For any $s \leq 0$, we have $v_s(x_1) = v_0(x_1)$, and furthermore, $v_0(x_1) < 0$ for any x_1 in the domain.

(v) If $x_1 < -\delta/2$ or $x_1 > 1 + \delta/2$, then $v_s(x_1) = v(x_1)$ for any s.

It is difficult to visualize the situation just from these conditions, but, as it is depicted in Figure 3.15, we perturb the function in such a way that it is the function $v(x_1)$ itself when the parameter s coincides with η, and we push down the values of the function $v(x_1)$ as s decreases to 0, so that it always takes negative values when $s = 0$.

In the neighborhood U, we perturb the vector field X to the following vector field $\tilde{X}$:

(3.113)
$$\tilde{X} = 2v_\rho(x_1)\frac{\partial}{\partial x_1} - 2x_2\frac{\partial}{\partial x_2} - \cdots - 2x_{\lambda+1}\frac{\partial}{\partial x_{\lambda+1}} + \cdots + 2x_m\frac{\partial}{\partial x_m},$$

where ρ in the first term is the function of class C^∞ on U defined by $\rho = x_2^2 + \cdots + x_m^2$, and $v_\rho(x_1)$ is obtained from the family $\{v_s(x_1)\}_s$ by substituting the function ρ for the parameter s. By the properties (iii) and (v) of the family $\{v_s(x_1)\}_s$, $\tilde{X}$ matches the original X on a subset of U distant from C. The second and later terms in (3.113) are not zero for the points not on the x_1-axis. On the x_1-axis, we also have $\tilde{X} \neq \mathbf{0}$, since the first term of $\tilde{X}$ satisfies $2v_0(x_1) < 0$ (as $\rho = 0$). Therefore $\tilde{X} \neq \mathbf{0}$ everywhere in U (cf. Figure 3.16). Since $\tilde{X}$ matches the original X on a subset of U distant from C, it can be extended smoothly to X outside of U, so that we obtain a vector field Y defined on all of M. The vector fields X and Y are distinct only in a small neighborhood of C.

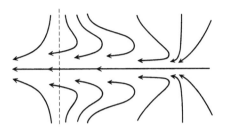

FIGURE 3.16. The vector field $\tilde{X}$

Consider $\overline{M}$ that contain two consecutive critical points p_{i-1} and p_i of the Morse function $f : M \to \mathbb{R}$. The vector field Y constructed above does not take the value $\mathbf{0}$. When we look closely at the construction of Y, we see that all integral curves of Y enter $\overline{M}$ through the lower boundary $\partial_0\overline{M} = f^{-1}(c_{i-2} + \varepsilon)$ and go out through the upper boundary $\partial_1\overline{M} = f^{-1}(c_i + \varepsilon)$. Using this fact, a function on $\overline{M}$

$$(3.114) \qquad\qquad \tilde{f} : \overline{M} \to \mathbb{R}$$

can be defined in such a way that it increases smoothly (in class C^∞) along the integral curves of Y from $c_{i-2} + \varepsilon$ to $c_i + \varepsilon$. Also $\tilde{f}$ can be constructed in such a way that it matches f on a neighborhood of the boundary. Then, we obtain a smooth function $g : M \to \mathbb{R}$ defined on all of M by extending $\tilde{f}$ to f outside of $\overline{M}$. This g is a desired function. It is almost obvious that g is a Morse function and Y is a gradient-like vector field for g. The properties (A), (B) of g can also be verified easily.

This completes the proof of Theorem 3.28. $\qquad\qquad\square$

As was shown in this proof, under the assumptions (i) and (ii) of Theorem 3.28, $\overline{M}$ is diffeomorphic to the direct product $\partial_0\overline{M} \times [0, 1]$, although $\overline{M}$ contains consecutive critical points p_{i-1} and p_i of f.

Rephrasing in terms of handlebodies. All the handlebodies we have considered so far are those associated with Morse functions $f : M \to \mathbb{R}$ on compact manifolds M without boundary. However, it can be shown that any handlebody (in general, any compact manifold N) is a subhandlebody associated with a Morse function $f : M \to \mathbb{R}$ on a certain compact manifold M without boundary. (We consider the so-called "double of N." The "double of N" is a smoothed manifold of the boundary $\partial(N \times I)$ of the direct product of N and the unit interval I. This is a closed manifold obtained from two copies of N by gluing them along ∂N.) From these facts, the results in Sections 3 and 4 can be stated as theorems about attaching handles, without mention of Morse functions. This is our goal of this subsection.

THEOREM 3.30 (Sliding handles: rephrased). *Suppose that a λ-handle $D^\lambda \times D^{m-\lambda}$ is attached by an attaching map $\psi : \partial D^\lambda \times D^{m-\lambda} \to \partial N$ on an m-dimensional manifold N with boundary, and suppose further that an isotopy $\{h_s\}_{s \in J}$ of ∂N is given. Here, we assume that $h_0 = \mathrm{id}$ and $h_1 = h$. Then the new handlebody $N \cup_{h \circ \psi} D^\lambda \times D^{m-\lambda}$ with the attaching map $h \circ \psi$, constructed by moving the "foot" $\partial D^\lambda \times D^{m-\lambda}$ of the handle by the isotopy, is diffeomorphic to the original handlebody $N \cup_\psi D^\lambda \times D^{m-\lambda}$ we had before the λ-handle was slid.*

It is convenient to define the term "belt sphere" here.

DEFINITION 3.31 (Belt sphere). Consider the situation where a λ-handle is attached to an m-dimensional manifold N with boundary. Set

$$N' = N \cup_\varphi D^\lambda \times D^{m-\lambda}.$$

In this case, the boundary $\mathbf{0} \times \partial D^{m-\lambda}$ of the co-core $\mathbf{0} \times D^{m-\lambda}$ of the λ-handle is called the *belt sphere* of this λ-handle. The belt sphere is an $(m-\lambda-1)$-dimensional sphere embedded in $\partial N'$ (cf. Figure 3.17).

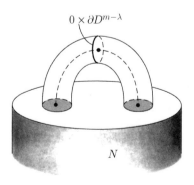

FIGURE 3.17. The belt sphere

A belt sphere rolls around the λ-handle as a belt rolls around the waist, which motivated its name. We rephrase Lemma 3.25 (separation of handles) in the preceding section using the belt sphere.

LEMMA 3.32 (Separation of handles: rephrased). *Suppose that a manifold N' is obtained from an m-dimensional manifold N with boundary by attaching a λ-handle, and suppose further that a manifold N'' is obtained from N' by attaching a μ-handle. If $\lambda \geq \mu$, then the attaching sphere of the μ-handle can be moved off of the belt sphere of the λ-handle, by moving the attaching sphere by an isotopy of $\partial N'$.*

Once the attaching sphere of the μ-handle is moved off of the belt sphere of the λ-handle, using arguments similar to those for Lemma 3.26 and Theorem 3.27, it is considered that the λ-handle and the μ-handle are attached to the original N simultaneously (and disjointly). Then, it can also be considered that the μ-handle with the smaller index was attached earlier.

Hence, we have proved the following lemma.

LEMMA 3.33 (Rearrangement of handles). *Any handlebody can be modified in such a way that the new one is described as follows. It is constructed first from a disjoint union of 0-handles, and then a disjoint union of 1-handles are attached on them, and then a disjoint union of 2-handles are attached, and so forth, so that handles are attached in ascending order of indices.*

THEOREM 3.34 (Canceling handles: rephrased). *Suppose that a manifold N' is obtained from an m-dimensional manifold N with boundary by attaching a λ-handle, and suppose further that a manifold N'' is obtained from N' by attaching a $(\lambda + 1)$-handle:*

$$N' = N \cup_\varphi D^\lambda \times D^{m-\lambda},$$
$$N'' = N' \cup_\psi D^{\lambda+1} \times D^{m-\lambda-1}.$$

If the belt sphere $\mathbf{0} \times \partial D^{m-\lambda}$ of the λ-handle and the attaching sphere $\partial D^{\lambda+1} \times \mathbf{0}$ of the $(\lambda+1)$-handle intersect transversely at a single point in the boundary $\partial N'$ of N', then N'' is diffeomorphic to N.

We look at examples for Theorem 3.34. Figure 3.18 illustrates the case where a 0-handle and a 1-handle are attached to N. In this case, the belt sphere of the 0-handle is a 2-dimensional sphere S^2 which is the surface of a 3-dimensional disk (a stuffed ball), and the attaching sphere of the 1-handle is two points, one of which lies on

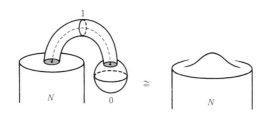

FIGURE 3.18. A 0-handle and a 1-handle are attached to N

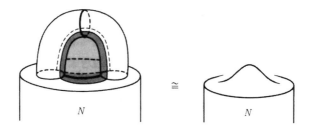

FIGURE 3.19. A 1-handle and a 2-handle are attached to N

the surface of the 0-handle. As we see from Figure 3.18, the union of the 0-handle and the 1-handle can be "swallowed" into N.

Next, Figure 3.19 illustrates a situation where a 1-handle and a 2-handle are attached. The attaching sphere (a circle) of the 2-handle and the belt sphere (which is also a circle) of the 1-handle intersect transversely at a single point. In this case, the union of the 1-handle and the 2-handle can be swallowed into N, as well.

Finally, we present a simple application of the theorems of sliding and canceling handles.

THEOREM 3.35. *Let M be a closed m-dimensional manifold. If M is connected, then there is a Morse function $f : M \to \mathbb{R}$ on M with only one critical point of index 0 and one critical point of index m.*

PROOF. By Theorem 3.22 (arranging critical points) and Theorem 3.27 (simultaneous handle attachment), for some Morse function $f : M \to \mathbb{R}$, we can assume that all the critical points of index 0 take

the same critical value c_0, all the critical points of index 1 take the
same critical value c_1 ($> c_0$), and so forth.

The handle decomposition with respect to such a Morse function
is of the form

(3.115)
$$M_{c_0+\varepsilon} = D_1^m \sqcup D_2^m \sqcup \cdots \sqcup D_r^m \quad \text{(a disjoint union of 0-handles)},$$

and $M_{c_1+\varepsilon}$ is obtained from $D_1^m \sqcup D_2^m \sqcup \cdots \sqcup D_r^m$ by attaching some
1-handles.

If $M_{c_1+\varepsilon}$ is not connected, then it consists of more than one connected component K_1, K_2, $\cdots$, K_s, but then, M itself would have
more than one component, since we attach handles of indices 2 or
higher to obtain M from $M_{c_1+\varepsilon}$. (Since the attaching sphere of a
handle of index 2 or higher is connected, it must be attached to a
single connected component. Hence the number of connected components does not decrease after attaching such handles.) Therefore,
$M_{c_1+\varepsilon}$ must be connected, as this contradicts the assumption that M
is connected.

Suppose there is more than one copy of D^m's in (3.115). From
the above consideration, they must be bridged together and become
connected as a whole. One D_r^m among those in (3.115) is bridged to
another D_i^m by a 1-handle $D^1 \times D^{m-1}$. Since the attaching sphere
$\partial D^1 \times \mathbf{0}$ ($= 2$ points) of this 1-handle intersects the belt sphere ∂D_r^m
of the 0-handle D_r^m at exactly one point, by Theorem 3.28, the crit-
ical point corresponding to the 0-handle D_r^m and the critical point
corresponding to the 1-handle $D^1 \times D^{m-1}$ are canceled out together.
Repeating this argument, we obtain a Morse function $g : M \to \mathbb{R}$
with only one critical point of index 0.

Next we consider the function with sign reversed, $-g : M \to \mathbb{R}$.
(This idea will become important in Section 3 of the next chapter.)
The function $-g$ is, so to speak, a Morse function obtained from g by
putting it upside-down. Although the sets of critical points are the
same for g and $-g$, if the index of a critical point p_i of g is λ, then
p_i, considered as a critical point of $-g$, has index $m - \lambda$.

In terms of handle decompositions, the roles of "core" and "co-
core" are interchanged, and a λ-handle of g becomes an $(m-\lambda)$-handle
of $-g$.

Using the above method, we perturb $-g$ and reduce the number
of 0-handles to 1. Then it can be considered that the number of m-
handles of g was reduced to 1. In this case, the number of m-handles
of $-g$, that is, the number of 0-handles of g, stays unchanged and is

1, so that the perturbed g has only one 0-handle and one m-handle. This completes the proof of Theorem 3.35.   □

If a closed m-dimensional manifold M is homotopy equivalent (which will be defined in Chapter 4, Section 1) to the m-dimensional sphere S^m, then M is called a *homotopy m-sphere*. The famous Poincaré conjecture states that a homotopy 3-sphere is homeomorphic to S^3. It has been almost 100 years since this conjecture was proposed by Poincaré in 1904, but the conjecture is still unsolved as of now.

In a paper published in 1961, however, Smale solved the generalized Poincaré conjecture for higher dimensions, [23]. Specifically, he proved that a homotopy m-sphere is homeomorphic to S^m for $m \geq 5$. Smale was awarded a Fields Medal in 1966 for this work.

The concept of handlebody was invented by Smale in [23]. He considered handle decompositions of M using Morse functions, and proved that all the handles can be canceled other than one 0-handle and one m-handle, under the hypothesis that M is a homotopy m-sphere with $m \geq 5$. Then, by Theorem 3.6, M is homeomorphic to S^m. This is the idea of Smale's proof.

The reason why the dimension must be higher is that the level surface $f^{-1}(c)$, which separates two critical points with indices differing by one, needs to have a large enough dimension so that the technique called Whitney's trick can be applied in the level surface, to pull apart the boundaries of the upper disk and the lower disk.

The generalized Poincaré conjecture for dimension 4 was solved by M.H. Freedman in 1981, [2]. Freedman was awarded a Fields Medal for this work.

Summary
3.1 A handle decomposition of a manifold can be obtained by a Morse function and a gradient-like vector field.
3.2 The number of handles and their indices are determined by a Morse function. Attaching maps are determined by a gradient-like vector field.
3.3 If N' is obtained from N by attaching a handle on a manifold N with boundary, then the diffeomorphism type of the manifold N' does not change if the attaching map is perturbed by an isotopy of the boundary ∂N.

3.4 When a handle of index less than or equal to λ is attached on a λ-handle, they can be pulled apart and made disjoint.

3.5 For an appropriate choice of a Morse function, the critical points can be arranged in such a way that the critical values increase as the indices increase.

3.6 If a $(\lambda + 1)$-handle is attached on a λ-handle, and furthermore if the attaching sphere of the $(\lambda + 1)$-handle intersects the belt sphere of the λ-handle transversely at a single point, then these two handles can be canceled together.

Exercises

3.1 Let $m \geq 1$. Prove that any homeomorphism $h : S^{m-1} \to S^{m-1}$ from the $(m - 1)$-dimensional sphere to itself can be extended to a homeomorphism $\overline{h} : D^m \to D^m$ from the m-dimensional disk to itself.

3.2 Let p_1 and p_2 be two points in the interior of the k-dimensional disk D^k. Construct an isotopy $\{h_t\}_{t \in J}$ of D^k such that $h_0 =$ id, $h_1(p_1) = p_2$, and h_t is constantly the identity map in a neighborhood of ∂D^k. (A bit difficult.)

3.3 In the interval $[c, d]$, two real numbers q_1 and q_2, $c < q_1 < q_2 < d$, are given, and also a and b are given in the same interval $[c, d]$ such that $c < a < d$ and $c < b < d$. For a small positive number δ, construct a C^∞-function

$$G : [c, d] \times (-\delta, 1 + \delta) \to [c, d]$$

with the following three properties. (Fairly difficult.)

(i) For a fixed s, $G(x, s)$ is a strictly increasing function in x, and $G(x, s)$ increases from c to d as x increases from c to d.

(ii) $G(q_1, 0) = a$, $G(q_2, 1) = b$.

(iii) For any $s \in [0, 1]$, $G(x, s) = x$ if x is in a small neighborhood of c or d. Also, $\dfrac{\partial}{\partial x} G(x, 0) = 1$ for x in a neighborhood of q_1, and $\dfrac{\partial}{\partial x} G(x, 1) = 1$ for x in a neighborhood of q_2.

3.4 Recall that for the Morse function $f = c_1 x_{11} + c_2 x_{22} + \cdots + c_m x_{mm}$ (where $1 < c_1 < c_2 < \cdots < c_m$) on $SO(m)$ given in Example 3.11, a diagonal matrix with $\varepsilon_1, \varepsilon_2, \cdots, \varepsilon_m$ on the diagonal is a critical point of f. Determine the index of the Hessian at this critical point.

CHAPTER 4

Homology of Manifolds

In this chapter, we discuss homology groups of manifolds using the results in previous chapters. Although we assume that the reader is familiar with homology groups, we review the material briefly in Section 1. (See [10], [21] for more details.) In Section 2 the Morse inequality is proved, which relates the number of critical points of a Morse function to the Betti number of a manifold. This is the most fundamental result in Morse theory. In Section 3, the Poincaré duality theorem is proved, which is a fundamental theorem in homology theory for manifolds. The intersection forms will be discussed in the last Section 4.

4.1. Homology groups

We start with a review of cell complexes. The interior $\mathrm{int}(D^i) = D^i - \partial D^i$ of the i-dimensional disk D^i (i-disk for short) is called an i-cell, and denoted by e^i. The i-disk D^i itself is called a *closed i-cell*, and denoted by $\bar{e}^i$. In the 0-dimensional case, e^0, as well as $\bar{e}^0$, is a single point. Therefore $e^0 = \bar{e}^0$.

DEFINITION 4.1 (Cell complex). A space obtained from 0-cells by attaching closed cells one after another, in ascending order of dimensions of cells, is called a (finite) *cell complex*. More precisely, it is defined inductively as follows.

(i) The disjoint union of finitely many points, $X^0 = \bar{e}_1^0 \sqcup \bar{e}_2^0 \sqcup \cdots \sqcup \bar{e}_{k_0}^0$, is a 0-dimensional cell complex.

(ii) Supose that a cell complex is defined for dimensions less than or equal to $i-1$, for $i \geq 1$. Let Y be a cell complex of dimension less than or equal to $i-1$. Consider a disjoint union of finitely many i-cells $\bar{e}_1^i \sqcup \bar{e}_2^i \sqcup \cdots \sqcup \bar{e}_{k_i}^i$ and let

$$h_i : \partial \bar{e}_1^i \sqcup \partial \bar{e}_2^i \sqcup \cdots \sqcup \partial \bar{e}_{k_i}^i \to Y$$

be a continuous map from its boundary to Y. The space

$$X^i = Y \cup_{h_i} (\bar{e}_1^i \sqcup \bar{e}_2^i \sqcup \cdots \sqcup \bar{e}_{k_i}^i)$$

obtained from Y by attaching $\bar{e}_1^i \sqcup \bar{e}_2^i \sqcup \cdots \sqcup \bar{e}_{k_i}^i$ with h_i is called an i-dimensional cell complex. The map h_i is called an *attaching map*.

This inductively defines an i-dimensional cell complex for all dimensions i.

Here, what we mean by attaching $\bar{e}_1^i \sqcup \bar{e}_2^i \sqcup \cdots \sqcup \bar{e}_{k_i}^i$ by h_i is to identify x and the point $h_i(x)$ of Y for each point x of $\partial \bar{e}_1^i \sqcup \partial \bar{e}_2^i \sqcup \cdots \sqcup \partial \bar{e}_{k_i}^i$.

In what follows, when we refer to a cell complex X, we mean a cell complex X^m of a certain dimension m. The union X^i of all the cells of dimensions less than or equal to i in X forms a cell complex, and is called the i-*skeleton* of X. The points of the 0-skeleton X^0 are sometimes called *vertices* of X.

It is easily seen that a cell complex X is a union of cells (without boundary), disjoint from each other.

EXAMPLE 4.2. There is a unique continuous map $h : \partial \bar{e}^m \to \bar{e}^0$ from the boundary $\partial \bar{e}^m$ of a closed m-dimensional cell to a single point $\bar{e}^0$, that is, the map collapsing $\partial \bar{e}^m$ to a single point $\bar{e}^0$. The m-dimensional cell complex

$$\bar{e}^0 \cup_h \bar{e}^m$$

is homeomorphic to the m-dimensional sphere S^m. Since an m-cell is obtained from S^m by deleting a single point e^0, it may be considered also that $S^m = e^0 \cup e^m$.

We review homology groups of cell complexes. Let X be a cell complex. Specify an "orientation" on each cell contained in X. Here, an "orientation" of a cell is defined as follows. If a cell e is of dimension q, then e can be identified with an open set of the q-dimensional Euclidean space, so that there is an array (i.e. a finite sequence) of vector fields

$$V = \langle \mathbf{v}_1, \mathbf{v}_2, \cdots, \mathbf{v}_q \rangle$$

such that they constitute an ordered basis of the tangent space at every point of e. Such an array is considered to determine an *orientation* of e. If there is a similar array

$$W = \langle \mathbf{w}_1, \mathbf{w}_2, \cdots, \mathbf{w}_q \rangle$$

on e, then we say that V and W determine the same orientation of e if the transformation matrix A from V to W has a positive determinant at every point of e. (If the determinant is positive at a single point of e, then it is positive at every point of e.) Otherwise we say that V and W determine the opposite orientations of e.

In particular, we observe the following. If σ is a permutation of $\{1, 2, \cdots, q\}$, then a necessary and sufficient condition for

$$\langle \mathbf{v}_{\sigma(1)}, \mathbf{v}_{\sigma(2)}, \cdots, \mathbf{v}_{\sigma(q)} \rangle \quad \text{and} \quad \langle \mathbf{v}_1, \mathbf{v}_2, \cdots, \mathbf{v}_q \rangle$$

to determine the same orientation of e is that σ is an even permutation.

Since we will need it later, we define more generally an "orientation" of a manifold.

DEFINITION 4.3 (Orientation of a manifold). Let M be an m-dimensional manifold (m-manifold for short). Let M be covered by finitely or infinitely many coordinate neighborhoods: $M = \bigcup_{\lambda \in \Lambda} U_\lambda$. For each coordinate neighborhood U_λ, an orientation is defined by means of an array of m linearly independent vector fields. Suppose now that orientations are given for all U_λ simultaneously in such a way that for any U_λ and U_μ, as long as the intersection $U_\lambda \cap U_\mu$ is not the empty set, the orientation of U_λ and that of U_μ coincide on $U_\lambda \cap U_\mu$. (That is, the determinant of the transformation matrix between two sets of m vector fields defined on these two sets is positive.) In this case, M is called *orientable*, and giving such orientations of U_λ simultaneously is called giving an *orientation* $\langle M \rangle$ on M, or *orienting* M.

If M is orientable and connected, then there are exactly two orientations on M; they are opposite to each other.

We go back to cell complexes. A cell with an orientation is denoted by $\langle e \rangle$. We fix an integer q (≥ 0) representing a dimension. Suppose that a cell complex X has a number k_q of q cells, each of which is given an arbitrary orientation:

$$\langle e_1^q \rangle, \ \langle e_2^q \rangle, \ \cdots, \ \langle e_{k_q}^q \rangle.$$

A formal sum with integer coefficients of these oriented cells

$$c = a_1 \langle e_1^q \rangle + a_2 \langle e_2^q \rangle + \cdots + a_{k_q} \langle e_{k_q}^q \rangle$$

is called a q-*chain* of X. The set consisting of all q-chains of X forms an abelian group $C_q(X)$, called the q-*dimensional chain group* of X.

Here, the addition is defined for

$$c = a_1 \langle e_1^q \rangle + a_2 \langle e_2^q \rangle + \cdots + a_{k_q} \langle e_{k_q}^q \rangle$$

and

$$d = b_1 \langle e_1^q \rangle + b_2 \langle e_2^q \rangle + \cdots + b_{k_q} \langle e_{k_q}^q \rangle$$

by

$$c + d = (a_1 + b_1) \langle e_1^q \rangle + (a_2 + b_2) \langle e_2^q \rangle + \cdots + (a_{k_q} + b_{k_q}) \langle e_{k_q}^q \rangle.$$

Subtraction is defined similarly.

Since $C_q(X)$ is a free abelian group generated by $\langle e_1^q \rangle, \langle e_2^q \rangle, \cdots,$ $\langle e_{k_q}^q \rangle$, we have

$$C_q(X) \cong \mathbb{Z} \oplus \mathbb{Z} \oplus \cdots \oplus \mathbb{Z} \quad (k_q \text{ copies of } \mathbb{Z}).$$

We also say that $\langle e^q \rangle$, a q-cell e^q with an orientation, and the same cell with the opposite orientation $\langle e^q \rangle'$, are the negatives of each other in $C_q(X)$: $\langle e^q \rangle' = -\langle e^q \rangle$.

Next we define the *boundary homomorphism*

$$\partial_q : C_q(X) \to C_{q-1}(X).$$

For this purpose we need to define the orientation $\langle S^{q-1} \rangle$ of the boundary S^{q-1} of a q-disk D^q naturally determined by a fixed orientation $\langle D^q \rangle$. First we consider the case $q \geq 2$. Let $\langle \mathbf{v}_1, \mathbf{v}_2, \cdots, \mathbf{v}_q \rangle$ be an array of q vector fields which determines the given orientation $\langle D^q \rangle$, and suppose that the first vector field $\mathbf{v}_1$ matches the outward radial directions in a neighborhood of a point p of S^{q-1}. In this situation, we consider that the rest $\langle \mathbf{v}_2, \cdots, \mathbf{v}_q \rangle$ determine an orientation on a coordinate neighborhood of p in S^{q-1}. The orientation $\langle S^{q-1} \rangle$ determined this way is the one naturally induced from $\langle D^q \rangle$.

In the case $q = 1$, D^1 is a line segment, and an orientation can be thought of as an arrow on the segment. The boundary of the segment consists of two points. Generally, a single point is considered to have the orientation $\langle p \rangle$, which is uniquely given and fixed. We define the induced orientations on the two end points e_1^0, e_2^0 of an oriented segment $\langle D^1 \rangle$ to be such that it coincides with the given orientation $\langle e_1^0 \rangle$ at the end point e_1^0 into which the arrow points, and is the opposite of the given orientation $\langle e_2^0 \rangle$ at the end point e_2^0 out of which the arrow points.

We go back to the definition of the boundary homomorphism $\partial_q : C_q(X) \to C_{q-1}(X)$. Let $\langle e_k^q \rangle$ be an oriented q-cell in X ($k = 1, 2, \cdots, k_q$). This cell with boundary, $\bar{e}_k^q$, is attached to the cell complex Y of dimension less than or equal to $q - 1$ by an attaching map $h : \partial \bar{e}_k^q \to Y$. Here $\partial \bar{e}_k^q$ is a $(q-1)$-dimensional sphere and the orientation $\langle \partial \bar{e}_k^q \rangle$ is the one naturally induced from the orientation $\langle e_k^q \rangle$ of e_k^q.

If X does not contain a $(q-1)$-cell, then $C_{q-1}(X) = \{0\}$, so that $\partial_q = 0$, obviously.

In what follows we assume that X contains at least one $(q-1)$-cell, and Y is the $(q-1)$-skeleton X^{q-1}. Let $\langle e_l^{q-1} \rangle$ be an oriented $(q-1)$-cell $(l = 1, 2, \cdots, k_{q-1})$; then the image of $\partial \bar{e}_k^q$ under the attaching map $h : \partial \bar{e}_k^q \to X^{q-1}$ wraps e_l^{q-1} some number of times. Here we call this number the "covering degree" temporarily. The covering degree is determined with the orientations of $\langle \partial \bar{e}_k^q \rangle$ and $\langle e_l^{q-1} \rangle$ taken into consideration. Such a covering degree is denoted by a_{kl}. Then the boundary homomorphism is defined by

$$\partial_q(\langle e_k^q \rangle) = a_{k1}\langle e_1^{q-1} \rangle + a_{k2}\langle e_2^{q-1} \rangle + \cdots + a_{kk_{q-1}}\langle e_{k_{q-1}}^{q-1} \rangle.$$

Since the chain group $C_q(X)$ is generated by $\langle e_1^q \rangle, \langle e_2^q \rangle, \cdots, \langle e_{k_q}^q \rangle$, this determines the boundary homomorphism.

We explain the covering degree in more detail. Since $\partial \bar{e}_k^q$ is diffeomorphic to the $(q-1)$-sphere S^{q-1}, the attaching map h of $\bar{e}_k^q$ is thought of as a map $h : S^{q-1} \to X^{q-1}$ from the sphere S^{q-1}.

Take a point p of e_l^{q-1} and a neighborhood U $(\subset e_l^{q-1})$, and perturb h continuously so that h becomes a map of class C^∞ on $h^{-1}(U)$. Furthermore, according to Sard's Theorem 2.23 in Chapter 2, we can assume, by rechoosing a nearby p in U if necessary, that p is not a critical value of $h|h^{-1}(U) : h^{-1}(U) \to U$. Then the inverse image $h^{-1}(p)$ consists of a set of finitely many points $\{q_1, q_2, \cdots, q_r\}$ in S^{q-1}.

In a neighborhood of each point q_i, consider an array $\langle \mathbf{v}_1, \mathbf{v}_2, \cdots, \mathbf{v}_{q-1} \rangle$ of vector fields tangent to S^{q-1} which gives the orientation $\langle S^{q-1} \rangle$ already fixed. We also pick and fix an array $\langle \mathbf{w}_1, \mathbf{w}_2, \cdots, \mathbf{w}_{q-1} \rangle$ which gives the orientation $\langle e_l^{q-1} \rangle$ of e_l^{q-1}. Using these vector fields, we express the derivative $dh_{q_i} : T_{q_i}(S^{q-1}) \to T_p(U)$ of $h|h^{-1}(U)$ at the point q_i in terms of matrices. Let $J_h(q_i)$ be the matrix thus obtained. Define $\varepsilon(q_i) = 1$ or $\varepsilon(q_i) = -1$ depending on whether the determinant $\det(J_h(q_i))$ is positive or negative, respectively.

In this situation, a precise definition of the covering degree is given by

$$a_{kl} = \varepsilon(q_1) + \varepsilon(q_2) + \cdots + \varepsilon(q_r).$$

It can be proved that the value of a_{kl} does not depend on the continuous perturbation of h, nor on the point $p \in e_l^{q-1}$ selected, using Sard's Theorem of codimension 1 (cf. [15]).

LEMMA 4.4.
$$\partial_{q-1} \circ \partial_q = 0, \quad \forall q.$$

As this is a well known lemma in homology theory, we omit the proof.

A sequence consisting of chain groups of a cell complex X and boundary homomorphisms

$$\cdots \to C_{q+1}(X) \xrightarrow{\partial_{q+1}} C_q(X) \xrightarrow{\partial_q} C_{q-1}(X) \xrightarrow{\partial_{q-1}}$$
$$\cdots \xrightarrow{\partial_2} C_1(X) \xrightarrow{\partial_1} C_0(X) \to \{0\}$$

is called the *chain complex* of X. For a fixed q, we set

$$
\begin{aligned}
Z_q(X) &= \operatorname{Ker} \partial_q := \{ \, c \in C_q(X) \mid \partial_q(c) = 0 \, \}, \\
B_q(X) &= \operatorname{Im} \partial_{q+1} := \{ \, c \in C_q(X) \mid c = \partial_{q+1}(c') \\
&\qquad\qquad\qquad \text{for some } c' \in C_{q+1}(X) \, \}
\end{aligned}
$$

and call $Z_q(X)$ the *q-dimensional cycle group*, $B_q(X)$ the *q-dimensional boundary group*. These are subgroups of $C_q(X)$, and by Lemma 4.4, we have
$$B_q(X) \subset Z_q(X) \subset C_q(X).$$
The elements of $Z_q(X)$ are called q-dimensional cycles of X, and those of $B_q(X)$ are called q-dimensional boundaries of X.

DEFINITION 4.5 (Homology group). We call the quotient group $Z_q(X)/B_q(X)$ the *q-dimensional homology group*, and denote it by $H_q(X)$:
$$H_q(X) = Z_q(X)/B_q(X).$$
The elements of the group $H_q(X)$ are called *homology classes*. The homology class to which a q-dimensional cycle z belongs is denoted by $[z]$.

EXAMPLE 4.6. For $m \geq 1$, the chain complex of the m-sphere $S^m = e^0 \cup e^m$ is of the form
$$\cdots \to \{0\} \to C_m(S^m) \to \cdots \to \{0\} \to C_0(S^m) \to \{0\}.$$
Here, $C_m(S^m) \cong C_0(S^m) \cong \mathbb{Z}$ and all the other groups $C_q(S^m)$ are $\{0\}$. We find that $\partial_m = 0$. In fact, if $m \geq 2$, then it is obvious

TABLE 4.1. Homology groups of S^m

q	0	1	$\cdots$	$m-1$	m
$H_q(S^m)$	$\mathbb{Z}$	$\{0\}$	$\cdots$	$\{0\}$	$\mathbb{Z}$

from the fact $C_{m-1}(S^m) = 0$, and if $m = 1$, then it follows from $\partial_1(\langle e^1 \rangle) = \langle e^0 \rangle - \langle e^0 \rangle = 0$. Therefore, we obtain

$$Z_m(S^m) \cong \mathbb{Z}, \quad B_m(S^m) \cong \{0\},$$

and the following result:

$$H_q(S^m) \cong \begin{cases} \mathbb{Z} & (q = m \text{ or } q = 0), \\ \{0\} & (\text{other } q). \end{cases}$$

This situation is represented in Table 4.1.

A continuous map between cell complexes

$$f : X \to Y$$

induces uniquely a homomorphism between homology groups

$$f_* : H_q(X) \to H_q(Y)$$

for each dimension q. Furthermore, the identity map $\mathrm{id}_X : X \to X$ corresponds to the identity homomorphism of the homology groups, and the composition $g \circ f$ of continuous maps corresponds to the composition $g_* \circ f_*$ of the homomorphisms.

The most important property of homology groups is the homotopy invariance, which we explain below.

DEFINITION 4.7 (Homotopy). Let X and Y be topological spaces, and $f, g : X \to Y$ continuous maps. We say that f is *homotopic* to g if there exists a continuous map $H : X \times [0, 1] \to Y$ from the direct product $X \times [0, 1]$ to Y such that, for all $x \in X$,

$$H(x, 0) = f(x), \quad H(x, 1) = g(x).$$

The continuous map H is sometimes called a *homotopy* from f to g.

The situation that f is homotopic to g is denoted by $f \simeq g$. As we easily see, if $f \simeq g$, then $g \simeq f$.

THEOREM 4.8. (Homotopy invariance of homology groups). *If two continuous maps $f, g : X \to Y$ are homotopic ($f \simeq g$), then these maps induce the identical homomorphism on homology groups:*

$$f_* = g_* : H_q(X) \to H_q(Y).$$

DEFINITION 4.9 (Homotopy equivalence). Let X and Y be two cell complexes. If there exist continuous maps $f : X \to Y$ and $g : Y \to X$ such that

$$g \circ f \simeq \mathrm{id}_X, \quad f \circ g \simeq \mathrm{id}_Y,$$

then X and Y are said to be *homotopy equivalent*, and we write $X \simeq Y$, or $f : X \simeq Y$.

The maps f and g are called *homotopy equivalences*, and f and g are sometimes called *homotopy inverses* of each other.

As a corollary of Theorem 4.8, we obtain

COROLLARY 4.10. *For cell complexes X and Y, if $f : X \simeq Y$, then*

$$f_* : H_q(X) \cong H_q(Y), \quad \forall q = 0, 1, 2, \cdots .$$

The homology groups $H_q(X)$ of a (finite) cell complex X are finitely generated abelian groups, so they have the form of

$$H_q(X) \cong \mathbb{Z} \oplus \mathbb{Z} \oplus \cdots \oplus \mathbb{Z} \oplus T,$$

where T is an abelian group containing only finitely many elements, and is called the *torsion part* of $H_q(X)$. The part $\mathbb{Z} \oplus \mathbb{Z} \oplus \cdots \oplus \mathbb{Z}$ is called the *free part*. The number of copies of $\mathbb{Z}$ in the free part (that is, the rank of $H_q(X)$) is called the *q-dimensional Betti number* of X, and is denoted by $b_q(X)$:

$$b_q(X) := \mathrm{rank} H_q(X), \quad q = 0, 1, 2, \cdots .$$

The Betti numbers play an important role in the next section.

REMARK. The torsion part T is a well-defined subgroup of $H_q(X)$, but the free part is not. Only its rank is well-defined.

THEOREM 4.11. (Euler-Poincaré formula) *Let X be an m-dimensional cell complex, and k_q the number of q-cells contained in X. Then we have*

$$\sum_{q=0}^{m}(-1)^q k_q = \sum_{q=0}^{m}(-1)^q b_q(X).$$

We refer to [10] or [21] for the proof (which is direct from the definitions). The value of this equality is denoted by $\chi(X)$ and is called the *Euler number*, or the *Euler-Poincaré characteristic*. The Euler number $\chi(X)$ is a homotopy invariant of X. For example, since the m-disk D^m is homotopy equivalent to a single point, we have $\chi(D^m) = \chi(\text{point}) = 1$.

EXAMPLE 4.12 (Euler numbers of spheres). Since the m-sphere S^m is decomposed to $e^0 \cup e^m$ as a cell complex, we have

$$\chi(S^m) = 1 + (-1)^m.$$

Namely, $\chi(S^m) = 2$ if m is even, and $\chi(S^m) = 0$ if m is odd.

PROPOSITION 4.13. *For cell complexes X, Y_1, Y_2, and Z, if $X = Y_1 \cup Y_2$ and $Z = Y_1 \cap Y_2$, then we have*

$$\chi(X) = \chi(Y_1) + \chi(Y_2) - \chi(Z),$$

where we assume that Y_1 and Y_2 are cell subcomplexes consisting of subsets of the set of cells of X, and Z is a cell subcomplex of Y_1 and Y_2.

For a proof, one counts the number of cells of each dimension, considering which of Y_1, Y_2 and Z they belong to.

4.2. Morse inequality

The goal of this section is to prove the following theorem.

THEOREM 4.14 (Morse inequality). *Let M be a closed m-manifold, and $f : M \to \mathbb{R}$ a Morse function on M. For the number k_λ of critical points of index λ and the λ-dimensional Betti number $b_\lambda(M)$ of M, the following inequality holds:*

$$k_\lambda \geq b_\lambda(M).$$

Since the Betti numbers $b_\lambda(M)$ are determined by the shape of M, we see from this inequality that the number of critical points of a Morse function on M is restricted by the shape of M. In particular, if $b_\lambda(M) > 0$, then a Morse function on M must have at least one critical point of index λ.

Morse theory investigates the relations between the shape of a manifold and functions on the manifold, and the Morse inequality is a typical result of Morse theory.

(a) Handlebodies and cell complexes

First, we clarify the relations between handlebodies and cell complexes. A handle decomposition gives rise to a structure of a cell complex, called a *cell decomposition* of a manifold. To state the conclusions, we define the notion of a mapping cylinder.

DEFINITION 4.15 (Mapping cylinder). For a given continuous map $h : K \to X$ between topological spaces, the space obtained from X by attaching $K \times [0,1]$,

$$X \cup_h K \times [0,1],$$

by identifying the point "at the bottom" $(x,0)$ of the direct product $K \times [0,1]$ and the point $h(x)$ of X for each point x of K, is called the *mapping cylinder* of h, and is denoted by M_h (cf. Figure 4.1).

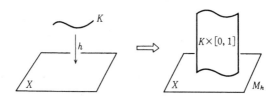

FIGURE 4.1. The mapping cyclinder M_h of $h : K \to X$

LEMMA 4.16. *The mapping cylinder M_h of $h : K \to X$ is homotopy equivalent to X. More precisely, the map $i : X \to M_h$, which naturally identifies X with a copy of X in M_h, is a homotopy equivalence.*

PROOF. We construct a continuous map $j : M_h \to X$ as follows. Looking at the structure $M_h = X \cup_h K \times [0,1]$, we set $j(p) = p$ for a point p of X, and set $j(x,t) = h(x)$ for a point (x,t) of $K \times [0,1]$. Then it is clear that a continuous map $j : M_h \to X$ from M_h to X is well-defined.

We show that $i : X \to M_h$ and $j : M_h \to X$ are homotopy inverses of each other. It is clear that

$$j \circ i = \mathrm{id}_X.$$

To show that $i \circ j \simeq \mathrm{id}_{M_h}$, we construct a homotopy $H : M_h \times [0,1] \to M_h$ from $i \circ j$ to id_{M_h} as follows. Namely, set $H(p,s) = p$ for a point p of X, and set

$$H((x,t),s) = (x,ts)$$

for a point (x,t) of $K \times [0,1]$. In this expression s is a parameter of homotopy, and (x,ts) in the right-hand side is a point of $K \times [0,1]$, and hence, a point of M_h.

It is easily seen that H gives the identity map on M_h for $s = 1$, and gives $i \circ j$ for $s = 0$, so that we have shown that

$$i \circ j \simeq \mathrm{id}_{M_h}.$$

In short, H is a continuous deformation which collapses the product part $K \times [0,1]$ in M_h down to $K \times \{0\}$ gradually.

Hence i gives a homotopy equivalence $X \simeq M_h$, and also we see that i and j are homotopy inverses of each other. $\square$

EXAMPLE 4.17. The m-disk D^m is homeomorphic to the mapping cylinder M_c of the map $c : S^{m-1} \to \{p\}$ which collapses the $(m-1)$-sphere S^{m-1} to a single point p.

In fact, regard D^m as the m-dimensional unit disk, and consider the map

$$D^m \to M_c$$

sending the center $\mathbf{0}$ of D^m to the point p in M_c, and sending the point x_t to the point (x, t) of M_c, where x_t is on the line segment from the center $\mathbf{0}$ of D^m to a point x on the boundary S^{m-1}, with the distance t from $\mathbf{0}$. Then this map defines a homeomorphism.

By Lemma 4.16, D^m is homotopy equivalent to a single point $\{p\}$.

Of course, this fact $D^m \simeq \{p\}$ can be proved easily without using Lemma 4.16.

THEOREM 4.18 (Handlebodies and cell complexes). *Let N be an m-dimensional handlebody. If the largest index of the handles contained in N is l, then N is homotopy equivalent to a certain l-dimensional cell complex X.*

More precisely;

(i) There exists a continuous map $h : \partial N \to X$ from the boundary ∂N of N to X such that N is homeomorphic to the mapping cylinder M_h of h. (Hence, by Lemma 4.16, we have $N \simeq X$.)

(ii) There is a one-to-one correspondence between the i-handles of N and the i-cells of X.

COROLLARY 4.19. *Let N be an m-dimensional handlebody. If k_i is the number of i-handles contained in N, then the Euler number of N is given by*

$$\chi(N) = \sum_{i=0}^{m} (-1)^i k_i.$$

In fact, this follows by combining Theorem 4.18 and the Euler-Poincaré formula.

We now prove Theorem 4.18.

PROOF. The idea is simple: we reduce the radii of the co-cores of handles smaller and smaller, and finally shrink the handles to their cores. The cell complex X is the one obtained by attaching the disks of cores of handles one after another. We explain the details below.

By Lemma 3.33 (rearrangement of handles: rephrased) in the preceding chapter, we can assume that N is a handlebody of the following form:

$$N = (h_1^0 \sqcup \cdots \sqcup h_{k_0}^0) \cup (h_1^1 \sqcup \cdots \sqcup h_{k_1}^1) \cup \cdots \cup (h_1^l \sqcup \cdots \sqcup h_{k_l}^l),$$

where h^λ represents a λ-handle $D^\lambda \times D^{m-\lambda}$. The notation h_i^λ denotes the i-th λ-handle.

Specifically, N is constructed from a disjoint union of 0-handles $h_1^0 \sqcup \cdots \sqcup h_{k_0}^0$ by attaching a disjoint union of 1-handles $h_1^1 \sqcup \cdots \sqcup h_{k_1}^1$, $\cdots$, and so forth.

Theorem 4.18 is proved by induction on the maximal index l of handles contained in N. If $l = 0$, then

$$N = D^m \sqcup \cdots \sqcup D^m \quad (k_0 \text{ copies of } m\text{-disks}).$$

If we regard the set of k_0 points $\{p_1, p_2, \cdots, p_{k_0}\}$ as a 0-dimensional cell complex X, then by Example 4.17, N is homeomorphic to the mapping cylinder M_c of the map $c : \partial N = \partial D^m \sqcup \partial D^m \sqcup \cdots \sqcup \partial D^m \to X = \{p_1, p_2, \cdots, p_{k_0}\}$ which collapses each sphere to a single point.

We assume that Theorem 4.18 is proved for handlebodies consisting of handles of indices less than or equal to $l - 1$, and prove it for a handlebody N whose maximal index of handles contained in N is l.

Let H be the subhandlebody of N consisting of all handles of indices less than l. Thus N is written as

$$N = H \cup_\psi (h_1^l \sqcup \cdots \sqcup h_{k_l}^l).$$

By the induction hypothesis, there are a cell complex Y and a continuous map $g : \partial H \to Y$ such that H is homeomorphic to the mapping cylinder M_g of g. For simplicity, we now assume that there is only one l-handle attached to H:

$$N = H \cup_\psi D^l \times D^{m-l}.$$

The handle $D^l \times D^{m-l}$ can be regarded as a mapping cylinder. Namely, if

$$c : D^l \times \partial D^{m-l} \to D^l \times \{\mathbf{0}\}$$

denotes the map which sends any $(x, y) \in D^l \times \partial D^{m-l}$ to $(x, \mathbf{0}) \in D^l \times \{\mathbf{0}\}$, then $D^l \times D^{m-l}$ is homeomorphic to the mapping cylinder M_c of c. By Example 4.17, regard D^{m-l} as a mapping cylinder of $\partial D^{m-l} \to \mathbf{0}$; then $D^l \times D^{m-l}$ can also be regarded as the product by D^l of this mapping cylinder.

Now we continue the proof of Theorem 4.18. Our strategy is to construct an l-dimensional cell complex X from Y and $D^l \times \{\mathbf{0}\}$, and construct a continuous map $h : \partial N \to X$ from g and c.

The attaching map $\psi : \partial D^l \times D^{m-l} \to \partial H$ is an embedding of class C^∞, so that $\partial D^l \times D^{m-l}$ can be regarded as a submanifold of ∂H via ψ. We make this identification in what follows.

Then, $\partial D^l \times \{\mathbf{0}\}$ is also a submanifold of ∂H. We denote the restriction of $g : \partial H \to Y$ on $\partial D^l \times \{\mathbf{0}\}$ by the same letter g. The l-dimensional cell complex X we desire is obtained from the cell complex Y by attaching a closed l-cell $D^l \times \{\mathbf{0}\}$ with the attaching map $g : \partial D^l \times \{\mathbf{0}\} \to Y$.

To simplify the notation, we set

$$K = \partial H - (\partial D^l \times \text{int} D^{m-l});$$

then the boundary ∂N of N can be decomposed (cf. Figure 4.2) as

$$(4.1) \qquad\qquad \partial N = K \cup D^l \times \partial D^{m-l}.$$

Our desired continuous map $h : \partial N \to X$ is defined on the portion $D^l \times \partial D^{m-l}$ by $h = c : D^l \times \partial D^{m-l} \to D^l \times \{\mathbf{0}\}$, where $D^l \times \{\mathbf{0}\}$ is regarded as an l-cell in X (attached to Y). Also, c is the map we used when the handle $D^l \times D^{m-l}$ is regarded as a mapping cylinder M_c.

To define h on K, we regard K as an $(m-1)$-manifold with boundary, and consider the collar neighborhood (here it is a closed collar neighborhood) V of the boundary $\partial K = \partial D^l \times \partial D^{m-l}$ in K. Thus we have

$$V \cong \partial K \times [0, 1],$$

and

$$\partial K = \partial K \times \{0\} = \partial D^l \times \partial D^{m-l}.$$

Let V_c be the mapping cylinder of the restriction $c|\partial K : \partial K = \partial D^l \times \partial D^{m-l} \to \partial D^l \times \{\mathbf{0}\}$ of the map c which we used to identify the handle with a mapping cylinder. Then V_c may be identified with $\partial D^l \times D^{m-l}$, and is homeomorphic to $\partial D^l \times D^{m-l} \cup V$ (which is $\partial D^l \times D^{m-l}$ with the collar V attached from outside). Let

$$j : V_c \to \partial D^l \times D^{m-l} \cup V \quad (\subset \partial H)$$

be this homeomorphism. Note that by mapping $V \cong \partial K \times [0,1]$ to $\partial K \times [0,1]$ inside the mapping cylinder $V_c = \partial D^l \times \{\mathbf{0}\} \cup \partial K \times [0,1]$, we have a natural map $i : V \to V_c$. On $\partial K \times \{1\}$ ($\subset \partial K \times [0,1] = V$) we have $j \circ i = $id, and on $\partial K \times \{0\}$ ($\subset \partial K \times [0,1] = V$) we have $j \circ i = c|\partial K$.

Now, a desired continuous map $h : \partial N \to X$ is defined on K as follows:

$$h(p) = \begin{cases} g \circ j \circ i(p) & \text{(if } p \in V), \\ g(p) & \text{(if } p \in \partial H - \partial K \times [0,1)). \end{cases}$$

Since two maps $g \circ j \circ i$ and g coincide on the intersection $\partial K \times \{1\}$ of the two regions in the right-hand side, a continuous map $h|K$ on K is well defined.

The map h defined on each portion of the decomposition (4.1) of ∂N coincides on the intersection $\partial D^l \times \partial D^{m-l}$, so we have obtained the desired continuous map $h : \partial N \to X$.

From the construction of h, it is clear that N is homeomorphic to the mapping cylinder M_h, and that there is a one-to-one correspondence between i-cells of the l-dimensional cell complex and i-handles of N.

This completes the proof of Theorem 4.18. $\square$

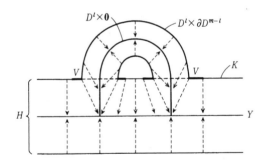

FIGURE 4.2. A construction of a continuous map $h : \partial N \to X$

REMARK. The cell complex X in the proof of Theorem 4.18 is considered to be embedded in the handlebody N from its construction.

In particular, if the handlebody N is a closed manifold, then N and X can be identified. For, if $\partial N = \emptyset$, then the mapping cylinder of the map $\emptyset \to X$ is nothing but X itself.

For example, as we have seen in Example 4.2, the m-sphere S^m itself can be regarded as an m-dimensional cell complex.

(b) Proof of the Morse inequality

We now prove Theorem 4.14. Let $f : M \to \mathbb{R}$ be a Morse function on a closed m-manifold M. Recall that we denote by k_λ the number of critical points of index λ.

Consider the handle decomposition defined by f. Then by Theorem 4.18 (handlebodies and cell complexes) and the above remark, M can be identified with a cell complex X, and there is a one-to-one correspondence between cells contained in X and the handles of M. In particular, the number of q-cells of X equals the number, k_q, of q-handles of M.

Consider the chain complex of X

$$\cdots \to C_q(X) \to C_{q-1}(X) \to \cdots \to C_1(X) \to C_0(X) \to \{0\}.$$

The rank of $C_q(X)$ is equal to the number k_q of q-cells of X for each q ($q = 0, 1, \cdots, m$).

Since the q-dimensional homology group $H_q(X)$ is obtained from a subgroup $Z_q(X)$ by taking the quotient by a smaller subgroup $B_q(X)$ of $Z_q(X)$, we have

$$k_q = \operatorname{rank} C_q(X) \geq \operatorname{rank} Z_q(X) \geq \operatorname{rank} H_q(X).$$

By identifying M and X we have

$$b_q(M) = b_q(X) = \operatorname{rank} H_q(X),$$

so that we obtain $k_q \geq b_q(M)$ ($q = 0, 1, \cdots, m$) from the above inequality. This proves the Morse inequality.

(c) Homology groups of complex projective space $\mathbb{C}P^m$

We determine the homology groups of the complex projective space $\mathbb{C}P^m$ of complex dimension m. We obtained a handle decomposition

$$\mathbb{C}P^m = h^0 \cup h^2 \cup \cdots \cup h^{2m}$$

of $\mathbb{C}P^m$ in Section 2 of the preceding chapter. To abbreviate notation, a λ-handle $D^\lambda \times D^{m-\lambda}$ is denoted by h^λ here. According to Theorem

4.18, $\mathbb{C}P^m$ can be represented as a cell complex from this handle decomposition:

$$\mathbb{C}P^m = e^0 \cup e^2 \cup \cdots \cup e^{2m}.$$

Hence the chain groups $C_q(\mathbb{C}P^m)$ of $\mathbb{C}P^m$ are given by

$$C_q(\mathbb{C}P^m) = \begin{cases} \mathbb{Z} & \text{(if } q \text{ is even, } 0 \leq q \leq 2m), \\ \{0\} & \text{(otherwise)}. \end{cases}$$

From this we see that the boundary homomorphisms are $\partial_q = 0$ $(\forall q)$, and the homology groups are computed as follows:

$$H_q(\mathbb{C}P^m) = \begin{cases} \mathbb{Z} & \text{(if } q \text{ is even, } 0 \leq q \leq 2m), \\ \{0\} & \text{(otherwise)}. \end{cases}$$

This is represented in Table 4.2.

TABLE 4.2. Homology groups of $\mathbb{C}P^m$

q	0	1	2	3	$\cdots$	$2m-1$	$2m$
$H_q(\mathbb{C}P^m)$	$\mathbb{Z}$	$\{0\}$	$\mathbb{Z}$	$\{0\}$	$\cdots$	$\{0\}$	$\mathbb{Z}$

4.3. Poincaré duality

In this section, first we introduce cohomology groups $H^q(X)$. For orientable closed manifolds, there is a nice relation called Poincaré duality between their homology and cohomology groups. The goal of this section is to prove this theorem.

(a) Cohomology groups

Consider the chain complex

$$\cdots \to C_{q+1}(X) \xrightarrow{\partial_{q+1}} C_q(X) \xrightarrow{\partial_q} C_{q-1}(X) \xrightarrow{\partial_{q-1}} \cdots \xrightarrow{\partial_1} C_0(X) \to \{0\}$$

of a cell complex X. The rank of $C_q(X)$ is equal to the number k_q of q-cells contained in X.

Let

$$C^q(X) = \{ f \mid f : C_q(X) \to \mathbb{Z} \}$$

be the set of all homomorphisms from $C_q(X)$ to $\mathbb{Z}$; then $C^q(X)$ becomes an abelian group with respect to the addition of homomorphisms. Here, the addition $f + g$ for $f, g \in C^q(X)$ is defined by

$$(f + g)(c) := f(c) + g(c), \quad \forall c \in C_q(X).$$

The subtraction is defined similarly.

The group $C^q(X)$ is called the *q-dimensional cochain group*, and each element f of $C^q(X)$ is called a *q-cochain*.

For any q-cochain $f : C_q(X) \to \mathbb{Z}$, the composition

$$f \circ \partial_{q+1} : C_{q+1}(X) \to \mathbb{Z},$$

with the boundary homomorphism $\partial_{q+1} : C_{q+1}(X) \to C_q(X)$, is a $(q+1)$-cochain. The map denoted by

$$\delta^q : C^q(X) \to C^{q+1}(X),$$

which assigns the $(q+1)$-cochain $f \circ \partial_{q+1}$ to a q-cochain f, is called the *coboundary homomorphism*. The coboundary homomorphism raises the dimensions of the cochain groups by one, contrary to the boundary homomorphism.

The sequence consisting of cochain groups and coboundary homomorphisms

(4.2)
$$\{0\} \to C^0(X) \xrightarrow{\delta^0} C^1(X) \to \cdots$$
$$\to C^{q-1}(X) \xrightarrow{\delta^{q-1}} C^q(X) \xrightarrow{\delta^q} C^{q+1}(X) \xrightarrow{\delta^{q+1}} \cdots$$

is called the *cochain complex* of X.

LEMMA 4.20.
$$\delta^{q+1} \circ \delta^q = 0, \quad \forall q.$$

PROOF. This can be proved easily from Lemma 4.4 and the definition of coboundary homomorphisms. □

The kernel $\mathrm{Ker}(\delta^q)$ of δ^q is denoted by $Z^q(X)$ and is called the *q-dimensional cocycle group*. This group $Z^q(X)$ is a subgroup of $C^q(X)$, and the elements of $Z^q(X)$ are called *cocycles*. Also, the image $\mathrm{Im}(\delta^{q-1})$ is denoted by $B^q(X)$ and is called the *q-dimensional coboundary group*, which is a subgroup of $C^q(X)$ as well. The elements of $B^q(X)$ are called *coboundaries*. By Lemma 4.20, we have

$$B^q(X) \subset Z^q(X) \subset C^q(X).$$

DEFINITION 4.21 (Cohomology group). We call the quotient group $Z^q(X)/B^q(X)$ the *q-dimensional cohomology group* and denote it by $H^q(X)$.

An element of $H^q(X)$ is called a *q-dimensional cohomology class*. The cohomology class to which a q-dimensional cocycle f belongs is denoted by $[f]$.

Since a q-cochain is a homomorphism $f : C_q(X) \to \mathbb{Z}$, for any q-cochain f and any q-chain c, an integer $f(c)$ is determined. By the definition of the coboundary homomorphism, we have

$$(\delta^q f)(c) = f(\partial_{q+1} c).$$

From this we can easily show that the value $f(c)$ for a q-dimensional cocycle f and a q-dimensional cycle c is determined only by the cohomology class $[f]$ and the homology class $[c]$. (In the next section on intersection forms, we will go over a similar argument in some detail, when we prove that an intersection number is determined between homology classes.) Therefore, a q-dimensional cohomology class $[f]$ determines a homomorphism $[f] : H_q(X) \to \mathbb{Z}$.

We set

$$\mathrm{Hom}(H_q(X), \mathbb{Z}) = \{ \phi \,|\, \phi : H_q(X) \to \mathbb{Z} \text{ is a homomorphism } \},$$

and then we obtain a map

$$\kappa : H^q(X) \to \mathrm{Hom}(H_q(X), \mathbb{Z})$$

by assigning $[f]$ considered as a homomorphism to a cohomology class $[f]$. This κ is a homomorphism. The next theorem is a special case of a weaker version of the theorem called the universal coefficient theorem (Theorem 4.4 in Chapter XII of [10] or Theorem 7.5 in [21]).

THEOREM 4.22. *The homomorphism*

$$\kappa : H^q(X) \to \mathrm{Hom}(H_q(X), \mathbb{Z})$$

is an "onto" map, and its kernel $\mathrm{Ker}(\kappa)$ *is the torsion part of* $H^q(X)$.

THEOREM 4.23 (Poincaré duality). *For an orientable closed m-manifold M, we have*

$$H^q(M) \cong H_{m-q}(M), \quad \forall q = 0, 1, \cdots, m.$$

(b) Proof of Poincaré duality

First, we make some observations on relations between boundary and coboundary homomorphisms.

If a cell complex X contains k_q q-cells $e_1^q, e_2^q, \cdots, e_{k_q}^q$, then the oriented cells $\langle e_1^q \rangle, \langle e_2^q \rangle, \cdots, \langle e_{k_q}^q \rangle$ can be taken as a natural basis of $C_q(X)$. A similar basis $\langle e_1^{q+1} \rangle, \langle e_2^{q+1} \rangle, \cdots, \langle e_{k_{q+1}}^{q+1} \rangle$ can be taken for $C_{q+1}(X)$, so that the boundary homomorphism $\partial_{q+1} : C_{q+1}(X) \to$

$C_q(X)$ is represented by a matrix using these bases. In fact, if

(4.3)
$$\partial_{q+1}(\langle e_i^{q+1}\rangle) = a_{i1}\langle e_1^q\rangle + a_{i2}\langle e_2^q\rangle + \cdots + a_{ik_q}\langle e_{k_q}^q\rangle,$$
$$i = 1, 2, \cdots, k_{q+1},$$

then ∂_{q+1} is represented by a $k_{q+1} \times k_q$ integral matrix

$$A = \begin{pmatrix} a_{11} & a_{12} & \cdots & a_{1k_q} \\ a_{21} & a_{22} & \cdots & a_{2k_q} \\ & & \cdots & \\ a_{k_{q+1}1} & a_{k_{q+1}2} & \cdots & a_{k_{q+1}k_q} \end{pmatrix}.$$

(In fact, it is more natural to consider the transpose ${}^tA$ of the matrix A, but here we keep it as is.)

For the q-dimensional cochain group $C^q(X)$, there is a basis $f_1^q, f_2^q, \cdots, f_{k_q}^q$, called the *dual basis* of $\langle e_1^q\rangle, \langle e_2^q\rangle, \cdots, \langle e_{k_q}^q\rangle$. Namely, f_j^q is a homomorphism $f_j^q : C_q(X) \to \mathbb{Z}$ defined by

(4.4)
$$f_j^q(\langle e_k^q\rangle) = \begin{cases} 1 & \text{(when } k = j), \\ 0 & \text{(when } k \neq j). \end{cases}$$

We represent the coboundary homomorphism $\delta^q : C^q(X) \to C^{q+1}(X)$ by a matrix using the dual basis $f_1^q, f_2^q, \cdots, f_{k_q}^q$ of $C^q(X)$ and the dual basis $f_1^{q+1}, f_2^{q+1}, \cdots, f_{k_{q+1}}^{q+1}$ of $C^{q+1}(X)$. Suppose the image of a dual basis element $f_j^q \in C^q(X)$ under $\delta^q : C^q(X) \to C^{q+1}(X)$ is expressed in $C^{q+1}(X)$ in terms of the dual basis $f_1^{q+1}, f_2^{q+1}, \cdots, f_{k_{q+1}}^{q+1}$ as

(4.5)
$$\delta^q(f_j^q) = b_{j1}f_1^{q+1} + b_{j2}f_2^{q+1} + \cdots + b_{jk_{q+1}}f_{k_{q+1}}^{q+1}.$$

The matrix B with b_{jk} at the (j, k)-entry is a matrix representing δ^q.

To compute the coefficients b_{jk}, we evaluate both sides of (4.5) at $\langle e_k^{q+1}\rangle$. By the defining equation (4.4) of the dual basis, the right-hand side can take a non-zero value only at the k-th term, and the value is equal to b_{jk}.

On the other hand, from the definition of the coboundary homomorphism and the matrix presentation of the boundary homomorphism, the left-hand side is equal to

(4.6)
$$\delta^q f_j^q(\langle e_k^{q+1}\rangle) = f_j^q(\partial_{q+1}\langle e_k^{q+1}\rangle)$$
$$= f_j^q(a_{k1}\langle e_1^q\rangle + a_{k2}\langle e_2^q\rangle + \cdots + a_{kk_q}\langle e_{k_q}^q\rangle)$$
$$= a_{kj},$$

and we obtain $b_{jk} = a_{kj}$. We state this fact as a lemma.

LEMMA 4.24. *The matrix B which represents $\delta^q : C^q(X) \rightarrow C^{q+1}(X)$ with respect to the dual basis is the transpose of the matrix A which represents $\partial_q : C_{q+1}(X) \rightarrow C_q(X)$.*

Now we start the proof of the Poincaré duality.

Let M be an oriented closed m-manifold. We use a handle decomposition

(4.7)
$$M = (h_1^0 \sqcup \cdots \sqcup h_{k_0}^0) \cup (h_1^1 \sqcup \cdots \sqcup h_{k_1}^1) \cup \cdots \cup (h_1^m \sqcup \cdots \sqcup h_{k_m}^m)$$

arranged in increasing order of indices. As we have shown in Theorem 4.18, M can be regarded naturally as an m-dimensional cell complex X, by regarding the cores of q-handles of this handle decomposition as q-cells. Choose and fix an arbitrary orientation $\langle e_j^q \rangle$ for the core e_j^q of the q-handle h_j^q. Then we obtain the chain complex

(4.8)
$$\cdots \rightarrow C_{q+1}(X) \xrightarrow{\partial_{q+1}} C_q(X) \rightarrow \cdots \rightarrow C_1(X) \xrightarrow{\partial_1} C_0(X) \rightarrow \{0\}$$

associated with the handle decomposition (4.7). As we explained in Section 1 of this chapter, the boundary homomorphism ∂_{q+1} is represented by the covering degree a_{jk} of how many times $\partial \overline{e}_j^{q+1}$ covers the q-cell $\overline{e}_k^q$ under the attaching map $h_j : \partial \overline{e}_j^{q+1} \rightarrow X^q$ of the cell complex, with the orientations taken into consideration. (Here X^q denotes the q-skeleton of M when regarded as a cell complex.)

When a handlebody is regarded as a cell complex, attaching maps of the cell complex are essentially the attaching maps of the handlebody (cf. Theorem 4.18). Using this fact, the covering degree a_{jk} can be interpreted in terms of a handle decomposition (4.7) as follows.

In the handle decomposition (4.7), let N^q be the subhandlebody with all the handles attached up to (and including) q-handles.

Suppose now that a $(q+1)$-handle $h_j^{q+1} = D^{q+1} \times D^{m-q-1}$ is attached to N^q by an attaching map $\varphi_j : \partial D^{q+1} \times D^{m-q-1} \rightarrow \partial N^q$. The image $\varphi_j(\partial D^{q+1} \times \mathbf{0})$ of the attaching sphere under φ_j is a q-sphere embedded in ∂N^q. To simplify the notation, we denote this by S_j^q in what follows:

$$S_j^q = \varphi_j(\partial D^{q+1} \times \mathbf{0}).$$

On the other hand, the belt sphere $\mathbf{0} \times \partial D^{m-q}$ of a q-handle $h_k^q = D^q \times D^{m-q}$ contained in N^q is an $(m-q-1)$-sphere, and also is a submanifold of ∂N^q. To simplify the notation again, denote this belt sphere by Σ_k^{m-q-1}:

$$\Sigma_k^{m-q-1} = \mathbf{0} \times \partial D^{m-q}.$$

The dimension q of the attaching sphere S_j^q and the dimension $m - q - 1$ of the belt sphere Σ_k^{m-q-1} add up to the dimension $m - 1$ of ∂N^q.

THEOREM 4.25 (General position: part 2). *Suppose that there are two closed submanifolds S_1, S_2 in a smooth manifold K, and that the sum of their dimensions is equal to that of K:*

$$\dim K = \dim S_1 + \dim S_2.$$

Then there exists an isotopy $\{h_t\}_{t \in J}$ of K with the following properties.

(i) $h_0 = \mathrm{id}_K$, and
(ii) $h_1(S_1)$ intersects S_2 transversely in finitely many points.

We omit the proof (see [14] or Chapter 2 of [4]).

Going back to the handlebody N^q, by this Theorem 4.25, we can find an isotopy $\{h_t\}_{t \in J}$ of the boundary ∂N^q such that $h_0 = \mathrm{id}_{\partial N^q}$ and $h_1(S_j^q)$ intersects the belt sphere Σ_k^{m-q-1} transversely at finitely many points. Using Theorem 3.21 (sliding handles), the attaching map φ_j can be replaced by $h_1 \circ \varphi_j$ by this isotopy, so that we can assume, from the beginning, that the attaching sphere S_j^q and the belt sphere Σ_k^{m-q-1} intersect transversely. (By applying this argument to the disjoint union of attaching spheres and the disjoint union of belt spheres, we can assume that any attaching sphere and any belt sphere intersect transversely in ∂N^q.)

First we consider a simple situation, where an attaching sphere S_j^q and a belt sphere Σ_k^{m-q-1} intersect at a single point in ∂N^q. In this case, as we see by making the q-handle h_k^q thinner and thinner to get a corresponding q-cell e_k^q, the attaching sphere S_j^q covers e_k^q exactly once, positively or negatively. In other words, we obtain $a_{jk} = \pm 1$ (cf. Figure 4.3, noting that $a_{jk} = 2$ in this figure).

To study general situations, we use the *intersection number* defined in the next Section 4. The intersection number is defined for two oriented submanifolds $\langle S_1 \rangle$ and $\langle S_2 \rangle$ intersecting transversely at finitely many points in an oriented manifold $\langle K \rangle$. (Here it is assumed that $\dim S_1 + \dim S_2 = \dim K$.)

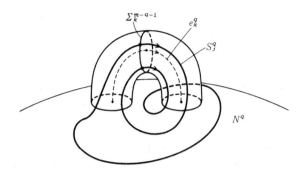

FIGURE 4.3. S_j^q covers e_k^q the same number of times
as S_j^q intersects Σ_k^{m-q-1}

The *sign of intersection*, $+1$ or -1, is assigned to each intersection
point of S_1 and S_2, and the intersection number between $\langle S_1 \rangle$ and
$\langle S_2 \rangle$,

$$\langle S_1 \rangle \cdot \langle S_2 \rangle,$$

is the integer obtained by adding up the signs of all intersection points.

Going back to a general situation where the attaching sphere S_j^q
intersects the belt sphere Σ_k^{m-q-1} many times, we note that S_j^q covers
e_k^q the same number of times as S_j^q intersects Σ_k^{m-q-1} with signs taken
into consideration, in this case. By choosing all the orientations of
$\langle S_j^q \rangle$, $\langle \Sigma_k^{m-q-1} \rangle$, and $\langle \partial N^q \rangle$ "naturally", one can compute that

$$(4.9) \qquad a_{jk} = (-1)^q \langle S_j^q \rangle \cdot \langle \Sigma_k^{m-q-1} \rangle.$$

The sign $(-1)^q$ in front is not essential here, so we omit details,
but we explain how to define "natural" orientations. Choose an arbi-
trary orientation $\langle e_k^q \rangle$ for the core e_k^q of a q-handle h_k^q. If we denote by
ε_k^{m-q} the co-core of h_k^q, then an orientation of ε_k^{m-q} is naturally determined from $\langle e_k^q \rangle$. Namely, since e_k^q and ε_k^{m-q} intersect transversely
at a single point, the orientation $\langle \varepsilon_k^{m-q} \rangle$ is determined in such a way
that

$$(4.10) \qquad \langle e_k^q \rangle \cdot \langle \varepsilon_k^{m-q} \rangle = 1,$$

under the assumption that an orientation $\langle M \rangle$ of the whole manifold
M is specified.

In general, an orientation of the boundary ∂N of a manifold N is
naturally induced from an orientation of N. (In Section 1 we discussed

this for disks, but a similar argument applies to general manifolds.) The orientations of attaching spheres S_j^q and belt spheres Σ_k^{m-q-1} are chosen to be the ones naturally induced from $\langle e_k^q \rangle$ and $\langle \varepsilon_k^{m-q} \rangle$, respectively, as their boundaries. We also choose the orientation of ∂N to be the one naturally induced from N. The equality (4.9) holds with these conventions.

The key point of the proof of Poincaré duality by Morse theory is to turn the manifold M "upside down." For this purpose, we just replace the original Morse function $f : M \to \mathbb{R}$ by $-f$,

$$-f : M \to \mathbb{R}.$$

As we have already seen in the proof of Theorem 3.35, the last theorem in Chapter 3, the set of critical points of the Morse function $-f$ is identical to the set of critical points of f. However, a critical point of index λ becomes a critical point of index $m - \lambda$ with respect to $-f$. This can be seen easily by looking at the standard form of f about a critical point. The roles of the core and co-core of a handle are switched, and a q-handle h_k^q transforms itself to an $(m - q)$-handle $h^*{}_k^{m-q}$.

The core of $h^*{}_k^{m-q}$ is ε_k^{m-q}, and the co-core is e_k^q.

The handle decomposition associated with $-f$ is

(4.11)
$$M = (h^*{}_1^0 \sqcup \cdots \sqcup h^*{}_{k_m}^0) \cup (h^*{}_1^1 \sqcup \cdots \sqcup h^*{}_{k_{m-1}}^1)$$
$$\cup \cdots \cup (h^*{}_1^m \sqcup \cdots \sqcup h^*{}_{k_0}^m)$$

where the numbers $k_m, k_{m-1}, \cdots, k_0$ of 0-handles, 1-handles, $\cdots$, m-handles, are, respectively, the numbers of m-handles, $(m-1)$-handles, $\cdots$, 0-handles before it was turned upside down.

Let X^* be the cell decomposition of M associated with the handle decomposition (4.11). From this cell complex, we obtain a chain complex

(4.12)
$$\cdots \to C_{m-q}(X^*) \xrightarrow{\partial^*_{m-q}} C_{m-q-1}(X^*) \to$$
$$\cdots \to C_1(X^*) \xrightarrow{\partial^*_1} C_0(X^*) \to \{0\}.$$

The basis $\langle \varepsilon_1^{m-q} \rangle, \langle \varepsilon_2^{m-q} \rangle, \cdots, \langle \varepsilon_{k_q}^{m-q} \rangle$ of $C_{m-q}(X^*)$ looks like the dual basis of $C^q(X)$, when we use the intersection numbers. In fact, if we compute the intersection number (in M) between a basis element

$\langle \varepsilon_k^{m-q} \rangle$ of $C_{m-q}(X^*)$ and a basis element $\langle e_j^q \rangle$ of $C_q(X)$, we obtain

$$\langle \varepsilon_k^{m-q} \rangle \cdot \langle e_j^q \rangle = \begin{cases} (-1)^{(m-q)q} & \text{(if } j = k), \\ 0 & \text{(if } j \neq k), \end{cases}$$

from the equality (4.10) and the formula (4.18) in the next section.

Furthermore, since $\text{rank}(C_{m-q}(X^*)) = \text{rank}(C^q(X)) = k_q$, an isomorphism

$$\Psi_{m-q} : C_{m-q}(X^*) \to C^q(X)$$

can be defined by assigning $(-1)^{(m-q)q} f_k^q$, the dual basis element f_k^q with sign adjusted, to a basis element $\langle \varepsilon_k^{m-q} \rangle$ of $C_{m-q}(X^*)$. Such an isomorphism Ψ_{m-q} is defined for each dimension q.

LEMMA 4.26. *For any q, the following diagram commutes up to sign:*

$$\begin{array}{ccc} C^q(X) & \xrightarrow{\ \delta^q\ } & C^{q+1}(X) \\ \Psi_{m-q} \uparrow & & \uparrow \Psi_{m-q-1} \\ C_{m-q}(X^*) & \xrightarrow[\ \partial_{m-q}^*\ ]{} & C_{m-q-1}(X^*) \end{array}$$

Namely, $\Psi_{m-q-1} \circ \partial_{m-q}^* = \pm\, \delta^q \circ \Psi_{m-q}$.

PROOF. We find a matrix presentation of the boundary homomorphism ∂_{m-q}^* with respect to the basis $\langle \varepsilon_k^{m-q} \rangle$ of $C_{m-q}(X^*)$ ($k = 1, \cdots, k_q$) and the basis $\langle \varepsilon_j^{m-q-1} \rangle$ of $C_{m-q-1}(X^*)$ ($j = 1, \cdots, k_{q+1}$). Set

(4.13)
$$\begin{aligned} &\partial_{m-q}^*(\langle \varepsilon_k^{m-q} \rangle) \\ &= c_{k1}\langle \varepsilon_1^{m-q-1} \rangle + \cdots + c_{kj}\langle \varepsilon_j^{m-q-1} \rangle + \cdots + c_{kk_{q+1}}\langle \varepsilon_{k_{q+1}}^{m-q-1} \rangle. \end{aligned}$$

Let N^{*m-q-1} be the subhandlebody consisting of all handles from 0-handles through $(m - q - 1)$-handles in the handle decomposition (4.11). We see that $N^{*m-q-1} \cup N^q = M$, and that the boundary ∂N^{*m-q-1} of N^{*m-q-1} is equal to the boundary ∂N^q of N^q. Considering the fact that the former belt sphere Σ_k^{m-q-1} is now the attaching sphere of an $(m - q)$-handle h^{*m-q}_k in ∂N^{*m-q-1}, and the former attaching sphere S_j^q is now the belt sphere of an $(m - q - 1)$-handle h^{*m-q-1}_j, we obtain

(4.14)
$$c_{kj} = (-1)^{m-q-1}\langle \Sigma_k^{m-q-1} \rangle' \cdot \langle S_j^q \rangle',$$

by an argument similar to the one when we obtained (4.9). This looks identical to (4.9), but we need to observe a few things here.

First of all, the intersection number is counted in $\langle \partial N^{*m-q-1} \rangle (= -\langle \partial N^q \rangle)$ in this case.

Also, to apply the same argument as when we computed a_{jk}, orientations of Σ_k^{m-q-1} and S_j^q must be defined in such a way that

$$\langle \varepsilon_j^{m-q-1} \rangle' \cdot \langle e_j^{q+1} \rangle' = 1, \quad \forall j.$$

This is why we have the symbol $'$ in the notations of orientations in (4.14). If we use the orientations $\langle \Sigma_k^{m-q-1} \rangle$ and $\langle S_j^q \rangle$ based on (4.10), then the difference in sign is $(-1)^{(m-q-1)(q+1)}$, from the formula (4.18) in the next section.

The above considerations imply that if we interchange the order of $\langle \Sigma_k^{m-q-1} \rangle$ and $\langle S_j^q \rangle$ using the formula (4.18), and compare with a_{jk} in the expression (4.9), then we obtain

$$(4.15) \qquad\qquad c_{kj} = (-1)^{q+1} a_{jk}.$$

In this formula, the sign $(-1)^{q+1}$ is not important for now. In any case, we have found that the matrix C representing ∂_{m-q}^* is equal to the transpose ${}^t\!A$ of the matrix A which represents ∂_{q+1} or its negative $-{}^t\!A$.

By Lemma 4.24, the matrix representing $\delta^q : C^q(X) \to C^{q+1}(X)$ with respect to the dual basis is the transpose of A. Hence it is now obvious that ∂_{m-q}^* and δ^q coincide up to sign by identifying $C_{m-q}(X^*)$ and $C^q(X)$ (resp. $C_{m-q-1}(X^*)$ and $C^{q+1}(X)$) by Ψ_{m-q} (resp. Ψ_{m-q-1}).

This completes the proof of Lemma 4.26. $\qquad\square$

If we use Lemma 4.26 and consider the isomorphism $\Psi_{m-q} : C_{m-q}(X^*) \to C^q(X)$ in each dimension q, we find that the chain complex

(4.16)

$$\cdots \to C_{m-q+1}(X^*) \xrightarrow{\partial_{m-q+1}^*} C_{m-q}(X^*) \xrightarrow{\partial_{m-q}^*} C_{m-q-1}(X^*) \to \cdots$$

is isomorphic to the cochain complex

$$(4.17) \qquad \cdots \to C^{q-1}(X) \xrightarrow{\pm\delta^{q-1}} C^q(X) \xrightarrow{\pm\delta_q} C^{q+1}(X) \to \cdots .$$

Therefore, the homology group $H_{m-q}(X^*)$ $(\cong H_{m-q}(M))$ of dimension $(m-q)$ of the chain complex (4.16) is isomorphic to the cohomology group $H^q(X)$ $(\cong H^q(M))$ of dimension q of the cochain complex (4.17). (The difference between δ^q or $-\delta^q$ in the coboundary homomorphism does not affect the structures of cohomology groups.) This proves the Poincaré duality $H_{m-q}(M) \cong H^q(M)$.

COROLLARY 4.27. *If M is an orientable, connected, closed m-manifold, then the highest dimensional homology and cohomology groups are isomorphic to $\mathbb{Z}$:*

$$H_m(M) \cong \mathbb{Z}, \quad H^m(M) \cong \mathbb{Z}.$$

PROOF. By Poincaré duality we have $H_m(M) \cong H^0(M)$ and $H^m(M) \cong H_0(M)$. On the other hand, for a connected cell complex X, it is easily proved from definitions that $H^0(M) \cong \mathbb{Z}$ and $H_0(M) \cong \mathbb{Z}$. □

Giving an orientation of M corresponds to specifying a generator of $H_m(M) \cong \mathbb{Z}$. Such a specified generator is denoted by $[M]$, and is called a *fundamental class* of M.

REMARK. For any non-orientable connected m-manifold M, it is known that $H_m(M) = \{0\}$.

COROLLARY 4.28 (Duality of Betti numbers). *If M is an orientable, connected, closed m-manifold, then the q-dimensional Betti number and the $(m-q)$-dimensional Betti number coincide:*

$$b_q(M) = b_{m-q}(M), \quad \forall q.$$

PROOF. By the universal coefficient theorem, we have rank $H^q(X) = $ rank $H_q(X)$. Combining with Poincaré duality, we obtain

$$b_q(M) = \text{rank } H_q(X) = \text{rank } H^q(X) = \text{rank } H_{m-q}(X) = b_{m-q}(M),$$

as desired. □

4.4. Intersection forms

Let M be an orientable m-manifold, and specify an orientation $\langle M \rangle$ of M. As we explained in Definition 4.3, locally $\langle M \rangle$ is given as an array of m vector fields, linearly independent at each point:

$$\langle V_1, V_2, \cdots, V_m \rangle.$$

Such an array of m vector fields is called an *m-frame of vector fields* associated with $\langle M \rangle$.

(a) Intersection numbers of submanifolds

Let S_1, S_2 be orientable submanifolds of M with dimensions p, q, respectively. Assume that $m = p + q$. Assume further that orientations $\langle S_1 \rangle$, $\langle S_2 \rangle$ are specified, and that S_1 and S_2 intersect in M transversely at finitely many points $\{q_1, q_2, \cdots, q_r\}$.

Consider one of the intersection points q_i. Choose a p-frame of vector fields $\langle V_1, V_2, \cdots, V_p \rangle$ on S_1 associated with $\langle S_1 \rangle$, and a q-frame of vector fields $\langle W_1, W_2, \cdots, W_q \rangle$ on S_2 associated with $\langle S_2 \rangle$, in a neighborhood of q_i.

DEFINITION 4.29 (Sign of intersection). The *sign of intersection* $\varepsilon(q_i)$ at the intersection q_i is defined as $\varepsilon(q_i) = +1$ or $\varepsilon(q_i) = -1$, depending on whether the array

$$\langle V_1, V_2, \cdots, V_p, W_1, W_2, \cdots, W_q \rangle,$$

formed by $\langle V_1, V_2, \cdots, V_p \rangle$ and $\langle W_1, W_2, \cdots, W_q \rangle$ in this order, matches an m-frame of vector fields associated with the orientation $\langle M \rangle$ of M, or the opposite one, respectively.

DEFINITION 4.30 (Intersection number). Let $S_1 \cap S_2 = \{q_1, q_2, \cdots, q_r\}$. Set

$$\langle S_1 \rangle \cdot \langle S_2 \rangle = \sum_{i=1}^{r} \varepsilon(q_i),$$

and call this number the *intersection number* between $\langle S_1 \rangle$ and $\langle S_2 \rangle$.

LEMMA 4.31. *When the order of $\langle S_1 \rangle$ and $\langle S_2 \rangle$ is interchanged, the following formula holds:*

(4.18) $$\langle S_1 \rangle \cdot \langle S_2 \rangle = (-1)^{pq} \langle S_2 \rangle \cdot \langle S_1 \rangle.$$

PROOF. Each time two vectors are interchanged in an m-frame of vector fields, the orientation determined by the m-frame is reversed. The m-frame of vector fields

$$\langle W_1, W_2, \cdots, W_q, V_1, V_2, \cdots, V_p \rangle$$

is obtained from $\langle V_1, V_2, \cdots, V_p, W_1, W_2, \cdots, W_q \rangle$ by performing interchanges of vectors pq times. Thus we obtain the formula (4.18). □

(b) Intersection forms

We generalize the intersection numbers between submanifolds to intersection numbers between homology classes. In what follows, let M be an oriented connected closed manifold, and consider the handle decomposition (4.7) associated with a Morse function $f : M \to \mathbb{R}$ and the handle decomposition (4.11) associated with $-f : M \to \mathbb{R}$. Using the same notation as in Section 3, let X, X^* be the corresponding cell complexes, respectively.

Recall that a q-cell e_k^q of X is the core of a q-handle h_k^q. Also, an $(m-q)$-cell ε_k^{m-q} of X^* is the co-core of h_k^q. The cells e_k^q and ε_k^{m-q} intersect transversely at a single point. Give an arbitrary orientation $\langle e_k^q \rangle$ on e_k^q, and give the orientation $\langle \varepsilon_k^{m-q} \rangle$ on ε_k^{m-q} in such a way that the intersection number is

$$(4.19) \qquad \langle e_k^q \rangle \cdot \langle \varepsilon_k^{m-q} \rangle = 1.$$

(See (4.10) of the preceding section.)

First, we define the intersection number between a q-chain of X and an $(m-q)$-chain of X^*. Let $c^q = \sum_{k=1}^{k_q} c_k \langle e_k^q \rangle$ $(c_k \in \mathbb{Z})$ be a q-chain of X, and $\gamma^{m-q} = \sum_{l=1}^{k_q} \gamma_l \langle \varepsilon_l^{m-q} \rangle$ $(\gamma_l \in \mathbb{Z})$ an $(m-q)$-chain of X^*.

DEFINITION 4.32 (Intersection number between chains). Define the *intersection number* between a q-chain c^q and an $(m-q)$-chain γ^{m-q} by

$$(4.20) \qquad c^q \cdot \gamma^{m-q} = \sum_{k,l=1}^{k_q} c_k \gamma_l \, \langle e_k^q \rangle \cdot \langle \varepsilon_l^{m-q} \rangle.$$

The q-cell e_k^q and the $(m-q)$-cell ε_l^{m-q} intersect transversely only when $k = l$, and do not intersect if $k \neq l$. Thus, the defining equality (4.20) is, in fact, equal to

$$c^q \cdot \gamma^{m-q} = \sum_{k=1}^{k_q} c_k \gamma_k,$$

where we used the equality (4.19).

By the defining equality (4.20), we get $(c + c') \cdot \gamma = c \cdot \gamma + c' \cdot \gamma$ and $c \cdot (\gamma + \gamma') = c \cdot \gamma + c \cdot \gamma'$, and the intersection number defines a bilinear form

$$\cdot : C_q(X) \times C_{m-q}(X^*) \to \mathbb{Z}$$

between $C_q(X)$ and $C_{m-q}(X^*)$.

We also see that $c^q \cdot \gamma^{m-q} = (-1)^{q(m-q)} \gamma^{m-q} \cdot c^q$, from the definition (4.20).

We now have the following key lemma.

LEMMA 4.33. *For all $c^{q+1} \in C_{q+1}(X)$ and all $\gamma^{m-q} \in C_{m-q}(X^*)$, we have*

$$(4.21) \qquad (\partial_{q+1}\, c^{q+1}) \cdot \gamma^{m-q} = (-1)^{q+1}\, c^{q+1} \cdot (\partial_{m-q}^*\, \gamma^{m-q}),$$

where

$$\partial_{q+1} : C_{q+1}(X) \to C_q(X) \quad and \quad \partial_{m-q}^* : C_{m-q}(X^*) \to C_{m-q-1}(X^*)$$

are boundary homomorphisms.

PROOF. It is sufficient to prove this lemma only for generators of $C_{q+1}(X)$ and those of $C_{m-q}(X^*)$. Hence we consider the case when

$$c^{q+1} = \langle e_j^{q+1} \rangle, \quad \gamma^{m-q} = \langle \varepsilon_k^{m-q} \rangle.$$

Let

$$(4.22) \qquad \begin{aligned} \partial_{q+1}(\langle e_j^{q+1} \rangle) &= \sum_{k=1}^{k_q} a_{jk} \langle e_k^q \rangle, \\ \partial_{m-q}^*(\langle \varepsilon_k^{m-q} \rangle) &= \sum_{j=1}^{k_{q+1}} c_{kj} \langle \varepsilon_j^{m-q-1} \rangle. \end{aligned}$$

Then the intersection number of the left-hand side of the equation (4.21) to be proved is $\partial_{q+1}(\langle e_j^{q+1} \rangle) \cdot \langle \varepsilon_k^{m-q} \rangle = a_{jk}$, and the intersection number of the right-hand side is $\langle e_j^{q+1} \rangle \cdot (\partial_{m-q}^* \langle \varepsilon_k^{m-q} \rangle) = c_{kj}$.

Since $a_{jk} = (-1)^{q+1} c_{jk}$ by the equality (4.15) of the preceding section, we obtain the desired formula (4.21). □

As a corollary of Lemma 4.33, we see that the intersection number between a q-dimensional boundary b^q of X and an $(m-q)$-dimensional cycle γ^{m-q} of X^* is zero. In fact, since a boundary b^q can be written as $b^q = \partial_{q+1}(c^{q+1})$ for a $(q+1)$-chain c^{q+1}, we obtain

$$(4.23) \qquad \begin{aligned} b^q \cdot \gamma^{m-q} &= (\partial_{q+1}(c^{q+1})) \cdot \gamma^{m-q} \\ &= (-1)^{q+1} c^{q+1} \cdot (\partial_{m-q}^* \gamma^{m-q}) \\ &= 0. \end{aligned}$$

Here, the next to last term is zero since it is assumed that γ^{m-q} is an $(m-q)$-dimensional cycle.

We see the following fact from this conclusion. In computing the intersection number

$$z^q \cdot \zeta^{m-q}$$

between a q-dimensional cycle z^q of X and an $(m - q)$-dimensional cycle ζ^{m-q} of X^*, the value of the intersection number stays unchanged after adding to z^q a q-dimensional boundary b^q. Similarly, the intersection number remains unchanged after adding to ζ^{m-q} an $(m - q)$-dimensional boundary β^{m-q}.

In other words, the intersection number

$$[z^q] \cdot [\zeta^{m-q}]$$

between a homology class $[z^q]$ of $H_q(X)$ and a homology class $[\zeta^{m-q}]$ of $H_{m-q}(X^*)$ is well-defined.

Since $H_q(X) \cong H_q(M)$ and $H_{m-q}(X^*) \cong H_{m-q}(M)$, we obtain a bilinear form

(4.24)
$$I : H_q(M) \times H_{m-q}(M) \to \mathbb{Z}, \quad \text{where} \quad I([z^q], [\zeta^{m-q}]) = [z^q] \cdot [\zeta^{m-q}],$$

by assigning the intersection numbers.

DEFINITION 4.34 (Intersection form). The bilinear form I is called the *intersection form*.

From the formula (4.18) we obtain

(4.25)
$$I(x, y) = (-1)^{q(m-q)} I(y, x), \quad \forall (x, y) \in H_q(M) \times H_{m-q}(M).$$

Since I is integer-valued, if $x \in H_q(M)$ is a *torsion* (i.e., if there is a non-zero integer n such that $nx = 0$), then $I(x, y) = 0$ for any $y \in H_{m-q}(M)$. The converse of this fact is also true.

LEMMA 4.35. *The intersection form is non-degenerate in the following sense.*

(i) For a fixed $x \in H_q(M)$, if $I(x, y) = 0$ for any $y \in H_{m-q}(M)$, then x is a torsion.

(ii) For any given homomorphism $\phi : H_{m-q}(M) \to \mathbb{Z}$, there exists an $x \in H_q(M)$ such that $I(x, y) = \phi(y), \forall y \in H_{m-q}(M)$.

The same holds if the roles of x and y are interchanged.

PROOF. This can be proved by combining Theorem 4.23 (Poincaré duality) and the universal coefficient theorem 4.22. It is seen that the composition

$$H_q(M) \cong H^{m-q}(M) \xrightarrow{\kappa} \mathrm{Hom}(H_{m-q}(M), \mathbb{Z})$$

is obtained by assigning the homomorphism $I(x, \cdot)$ (or its negative) to each $x \in H_q(M)$ by going back to the proof of Poincaré duality and the definition of κ. $\square$

If M is even-dimensional ($\dim(M) = 2n$), then since $m - n = n$, the intersection form is defined on the middle-dimensional $H_n(M)$:

$$I : H_n(M^{2n}) \times H_n(M^{2n}) \to \mathbb{Z}.$$

As a special case of the formula (4.25), we obtain

$$(4.26) \qquad I(x,y) = (-1)^n I(y,x), \quad \forall x,y \in H_n(M^{2n}).$$

Furthermore, if n is even (so that $m = 4k$ is a multiple of 4), then we see from this formula that I is symmetric ($I(x,y) = I(y,x)$). Hence we obtained the following theorem.

THEOREM 4.36. *If the dimension of M is a multiple of 4 ($\dim(M) = 4k$), then the intersection form*

$$I : H_{2k}(M) \times H_{2k}(M) \to \mathbb{Z}$$

is a non-degenerate symmetric bilinear form.

(c) Intersection numbers of submanifolds and intersection forms

At the beginning of this section we defined the intersection number $\langle S_1 \rangle \cdot \langle S_2 \rangle$ for oriented submanifolds S_1, S_2 of dimensions p, q, respectively, in an oriented m-manifold M. (It was, however, assumed that $\dim M = \dim S_1 + \dim S_2$, and that S_1 and S_2 intersect transversely at finitely many points.)

If S_1 and S_2 are closed submanifolds, then the fundamental classes $[S_1] \in H_p(S_1)$ and $[S_2] \in H_q(S_2)$ are determined, by Corollary 4.27 of Poincaré duality. Regard these as elements of $H_p(M)$ and $H_q(M)$ by mapping them by homomorphisms between homology groups induced from inclusion maps $j_1 : S_1 \to M$ and $j_2 : S_2 \to M$. Then the intersection number $I([S_1], [S_2])$ is determined by the intersection form I.

The following theorem looks obvious at first sight, but in fact, the proof is not so trivial.

THEOREM 4.37. $\langle S_1 \rangle \cdot \langle S_2 \rangle = I([S_1], [S_2])$.

The reason why the proof is not easy is that our definition of the intersection form started with the intersection numbers between chains of specific forms $\sum_{k=1}^{k_p} c_k \langle e_k^p \rangle$ and $\sum_{l=1}^{k_p} \gamma_l \langle \varepsilon_l^{m-p} \rangle$. To relate the intersection number $\langle S_1 \rangle \cdot \langle S_2 \rangle$ between submanifolds to $I([S_1], [S_2])$, we have to move S_1 and S_2 by smooth homotopies and put them into subhandlebodies N^p and $N^{*\ (m-p)}$ respectively. The proof that the intersection number remains unchanged before and after such homotopies becomes cumbersome at the homology-theoretic level.

The easiest proof, probably, is to regard the above smooth homotopies as maps from the products $S_1 \times [0,1]$ and $S_2 \times [0,1]$ to $M \times [0,1]$ and make them "transverse" to each other. Then the intersection between homotopies becomes finitely many curves, and the transverse intersection points before/after the move appear as their boundary points. This way the number of intersection points is proved to be unchanged before/after the move, using the curves of intersections. Details are omitted (see Chapter 3 of [4]).

If we use the fact that the intersection form I is defined for homology classes to prove Theorem 4.37, then we fall into a sort of circular reasoning.

The intersection form I was defined under the assumption that M is closed, but according to Theorem 4.37, the intersection form I can be defined "locally," so to speak, by counting the intersection numbers $\langle S_1 \rangle \cdot \langle S_2 \rangle$. Then the intersection form can be defined for manifolds with boundaries (in this case, however, I is not necessarily non-degenerate), and we see that Theorem 4.37 is not an obvious fact from this point of view either.

Summary

4.1 (Morse inequality) The number k_λ of critical points of index λ of a Morse function on M is greater than or equal to the Betti number $b_\lambda(M)$.

4.2 (Poincaré duality) If M is an orientable closed m-manifold, then $H^q(M) \cong H_{m-q}(M)$ for all q.

4.3 If M is an orientable closed m-manifold, then the intersection form $I : H^q(M) \times H_{m-q}(M) \to \mathbb{Z}$ is defined for all q. The intersection form is a non-degenerate bilinear form. It has a symmetry: $I(x,y) = (-1)^{q(m-q)} I(y,x)$.

4.4 The intersection form I can be defined for manifolds with boundary, but in that case it is not necessarily non-degenerate.

Exercises

4.1 Determine the homology groups of the projective plane P^2 using the handle decomposition of P^2 obtained in Section 2 of the preceding chapter.

4.2 Recall that the 2-dimensional homology group of the complex projective plane $\mathbb{C}P^2$ is $H_2(\mathbb{C}P^2) \cong \mathbb{Z}$. Let x be a generator of $H_2(\mathbb{C}P^2)$, and let $I : H_2(\mathbb{C}P^2) \times H_2(\mathbb{C}P^2) \to \mathbb{Z}$ be the intersection form. Prove that $I(x,x) = \pm 1$. (In fact, it can be shown that $I(x,x) = 1$ with the natural orientation of $\mathbb{C}P^2$ defined from its complex structure.)

4.3 If M is an orientable closed manifold of an odd dimension, then its Euler number $\chi(M)$ is 0.

CHAPTER 5

Low-dimensional Manifolds

Manifolds of dimension less than or equal to 4 are called low-dimensional manifolds. The theory of low-dimensional manifolds is deeply related to handlebody theory. In this chapter, we introduce fundamental aspects of low-dimensional manifolds from the point of view of handlebodies. First, we briefly review fundamental groups. See [10] for more details.

5.1. Fundamental groups

Let X be a connected cell complex. Pick and fix a point x_0 in X and call it the *base point*. Often a point of the 0-skeleton X^0 (i.e. a vertex) is chosen as the base point. A *loop* in (X, x_0) is a continuous map

$$f : (I, \partial I) \to (X, x_0)$$

from the interval $I = [0, 1]$ to X such that f maps both end points $\partial I = \{0, 1\}$ of I to x_0.

Two loops f, g are defined to be *homotopic* (with both endpoints ∂I of I fixed at the base point x_0) if there exists a continuous map

$$H : (I \times [0, 1], \partial I \times [0, 1]) \to (X, x_0)$$

such that

$$f(t) = H(t, 0), \quad g(t) = H(t, 1), \quad \forall t \in I.$$

In this case, we write $f \simeq g$.

The relation $\simeq$ is an equivalence relation in the set of all loops in (X, x_0), and the set of equivalence classes (homotopy classes) of this relation is denoted by

$$\pi_1(X, x_0).$$

The homotopy class to which a loop f belongs is written as $\{f\}$. The necessary and sufficient condition for

$$\{f\} = \{g\}$$

in $\pi_1(X, x_0)$ is that $f \simeq g$.

The set $\pi_1(X, x_0)$ becomes a group under the product of loops. The product $f \cdot g$ of two loops f, g is the loop defined by

$$(5.1) \qquad f \cdot g(t) = \begin{cases} f(2t) & \text{(for } 0 \le t \le \frac{1}{2}), \\ g(2t - 1) & \text{(for } \frac{1}{2} \le t \le 1). \end{cases}$$

This loop goes around f first, and then goes around g.

The homotopy class $\{f \cdot g\}$ to which $f \cdot g$ belongs is determined by the homotopy classes $\{f\}$, $\{g\}$ of f, g. Thus, by defining the product $\{f\} \cdot \{g\}$ between $\{f\}$ and $\{g\}$ to be $\{f \cdot g\}$, a "product" (multiplication) is defined on $\pi_1(X, x_0)$. The set $\pi_1(X, x_0)$ becomes a group with respect to this product. The identity element of this group is the homotopy class to which the loop f_0 defined by $f_0(I) = x_0$ belongs. Here f_0 is the trivial loop which stays fixed at the base point x_0.

DEFINITION 5.1 (Fundamental group). The group $\pi_1(X, x_0)$ is called the *fundamental group* of (X, x_0).

If X is connected, it can be proved that $\pi_1(X, x_0) \cong \pi_1(X, y_0)$ if the base point x_0 is replaced by another point y_0. So the choice of the base point is not essential, here.

For a continuous map $h : (X, x_0) \to (Y, y_0)$, a loop in (Y, y_0) is obtained by mapping a loop in (X, x_0) by h. This way a homomorphism

$$h_* : \pi_1(X, x_0) \to \pi_1(Y, y_0)$$

is induced.

LEMMA 5.2 (Homotopy invariance). *If $h : X \to Y$ is a homotopy equivalence and $h(x_0) = y_0$, then $h_* : \pi_1(X, x_0) \to \pi_1(Y, y_0)$ is an isomorphism.*

There is no mention of the base points in Definition 4.9 of homotopy equivalence in Chapter 4, and for this reason, the proof of this lemma is not as easy as it looks (see Theorem 8.3 in Chapter II of [10]).

EXAMPLE 5.3 (Fundamental group of the circle S^1). Take some point x_0 of the circle S^1 as the base point. Then it is known that

$$\pi_1(S^1, x_0) \cong \mathbb{Z},$$

where $\mathbb{Z}$ is given the group structure with addition as operation. If we construct a map deg : $\pi_1(S^1, x_0) \to \mathbb{Z}$ by assigning the number

$(\deg(f))$ of times a given loop f, with the base point x_0, wraps around S^1, then

$$\deg(f \cdot g) = \deg(f) + \deg(g).$$

This map gives an isomorphism.

EXAMPLE 5.4 (Fundamental group of spheres). Let S^n be a sphere of dimension at least 2, and take the base point x_0 on it. Then

$$\pi_1(S^n, x_0) \cong \{1\} \quad \text{(the trivial group)}.$$

This is proved as follows. Let f be an arbitrary loop with base point x_0. If f is onto, then move f by a homotopy to make the image a polygonal loop f'. Since $n \geq 2$, the image misses some point y_0. Then f' is contained in $S^n - \{y_0\}$, which is homeomorphic to an n-cell e^n, and can be contracted to the base point x_0 by a continuous deformation. Therefore, f' can contracted to the base point x_0 as well. Since this means that all loops are homotopic to the trivial loop, it is proved that $\pi_1(S^n, x_0)$ is the trivial group consisting only of the identity element 1.

If $\pi_1(X, x_0) \cong \{1\}$, the space X is called *simply connected*. A sphere S^n of dimension greater than or equal to 2 is simply connected.

Although in the above two examples, the groups $\pi_1(X, x_0)$ are abelian, in general, fundamental groups $\pi_1(X, x_0)$ are non-abelian.

EXAMPLE 5.5 (Bouquet of circles). The cell complex obtained from k copies of the circle $S_1^1, S_2^1, \cdots, S_k^1$ by attaching (gluing) them at a single point x_0 is called a *bouquet* of k circles. It is denoted by $S_1^1 \vee S_2^1 \vee \cdots \vee S_k^1$. For this space, it is shown that

$$\pi_1(S_1^1 \vee S_2^1 \vee \cdots \vee S_k^1, x_0) \cong F_k \quad \text{(free group of rank } k),$$

see Figure 5.1.

The free group F_k of rank $k \geq 2$ is non-abelian. Let us review the definition of F_k.

Take k letters $x_1, x_2, \cdots, x_k$ as generators. Elements of F_k are written as products of powers (with negative exponents allowed) of generators formed in any order, such as

$$x_2 x_1^{-1} x_3^4, \quad x_3^{-4} x_1^3 x_2^{-6}, \quad x_2^4, \quad x_1^{-1}.$$

Thus, the elements of F_k are *words* in the alphabet

$$\{x_1, x_1^{-1}, x_2, x_2^{-1}, \cdots, x_k, x_k^{-1}\};$$

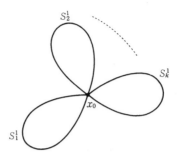

FIGURE 5.1. Bouquet of circles

the empty word, 1, is also added. Products in distinct orders are distinct elements of F_k. For example, $x_1 x_2 \neq x_2 x_1$.

For two elements of F_k, a product is defined by operation of "concatenation." We apply the axiom of exponents only when two powers of the same generator are adjacent. The 0-th power of any generator is identified with the identity element 1.

For example, the product of $x_2 x_1^{-1} x_3^4$ and $x_3^{-4} x_1^3 x_2^{-6}$ is computed as

$$
\begin{aligned}
(x_2 x_1^{-1} x_3^4) \cdot (x_3^{-4} x_1^3 x_2^{-6}) &= x_2 x_1^{-1} x_3^4 x_3^{-4} x_1^3 x_2^{-6} \\
&= x_2 x_1^{-1} x_3^0 x_1^3 x_2^{-6} \\
&= x_2 x_1^{-1} x_1^3 x_2^{-6} \\
&= x_2 x_1^2 x_2^{-6}.
\end{aligned}
$$

(5.2)

The free group F_k with generators $x_1, x_2, \cdots, x_k$ is sometimes denoted by

$$\langle x_1, x_2, \cdots, x_k \rangle.$$

If a loop f in $S_1^1 \vee S_2^1 \vee \cdots \vee S_k^1$, after departing from the base point, goes three times around S_2^1, say, and then goes (-2) times around S_1^1, then we assign the element $x_2^3 x_1^{-2}$ of F_k to $\{f\}$. Such an assignment $\pi_1(S_1^1 \vee S_2^1 \vee \cdots \vee S_k^1) \to F_k$ is the isomorphism in Example 5.5.

For computations of fundamental groups, van Kampen's theorem is useful. The following theorem is a special case.

THEOREM 5.6 (Special case of van Kampen's theorem). *Let $n \geq 2$. For a given continuous map $h : \partial D^n \to X$ from the boundary of the n-disk to X, consider the space $X \cup_h D^n$ obtained by attaching*

D^n to X by h. *Take the base point x_0 of X. Then the homomorphism $i_* : \pi_1(X, x_0) \to \pi_1(X \cup_h D^n, x_0)$ induced from the inclusion $i : (X, x_0) \to (X \cup_h D^n, x_0)$ is*

   (i) *an isomorphism if $n \geq 3$, and*
   (ii) *i_* is an "onto" homomorphism, with its kernel $\mathrm{Ker}(i_*)$ the normal subgroup $N(\{h\})$ of $\pi_1(X, x_0)$ generated by $\{h\}$, where $h : \partial D^2 = S^1 \to X$ is regarded as a loop.*

Refer to §4 in Chapter IV of [10] for a proof.

A caution should be paid to (ii) in this theorem. That is, the map $h : \partial D^2 \to X$ is a continuous map, but the image $h(S^1)$ does not necessarily pass through x_0, so h may not be a loop with x_0 as its base point.

Therefore, to state (ii) more precisely, we have to consider a loop h connected to x_0. For this purpose, take a point z_0 on S^1, and connect the base point x_0 to $h(z_0)$ by a curve C in X. Then consider the loop h' which starts from x_0, goes along C to $h(z_0)$, goes around the loop h once and comes back to $h(z_0)$, and then goes back along C to x_0. The loop h' is a loop in X with the base point x_0, so it is considered that $\{h'\} \in \pi_1(X, x_0)$, and the normal subgroup generated by $\{h'\}$ is the subgroup $N(h)$.

Although the choice of the curve C is not unique, the difference in $\{h'\}$ caused by taking a different C' is $\{h\}$ up to conjugacy in $\pi_1(X, x_0)$, so that the resulting normal subgroup $N(h)$ does not depend on the choice of C.

The above theorem (ii) can be restated using the homomorphism theorem as

$$\pi_1(X \cup_h D^2, x_0) \cong \pi_1(X, x_0)/N(h).$$

The right-hand side is the group obtained from $\pi_1(X, x_0)$ by adding the relation $h = 1$ (that h is equal to the identity element).

A general method of computing the fundamental groups of cell complexes is obtained by using van Kampen's Theorem 5.6. Let X be a connected cell complex, and take a vertex x_0 as the base point.

The 1-skeleton X^1 is homotopy equivalent to the bouquet of some number (say, k copies) of circles. Hence

$$\pi_1(X^1, x_0) \cong \langle x_1, x_2, \cdots, x_k \rangle.$$

The 2-skeleton X^2 is obtained from X^1 by attaching some number (say, m copies) of 2-cells to it. The boundary of a cell $\bar{e}_i^2$ is regarded as a loop in X^1, so that the loop connected to x_0 is written as a product of powers of generators $x_1, x_2, \cdots, x_k$. Denote this product by r_i for simplicity. By applying van Kampen's theorem m times, we obtain the following theorem.

THEOREM 5.7. (Fundamental groups of cell complexes). *The fundamental group $\pi_1(X, x_0)$ is isomorphic to the quotient group of the free group $\langle x_1, x_2, \cdots, x_k \rangle$ by the normal subgroup $N(r_1, r_2, \cdots, r_m)$ generated by $r_1, r_2, \cdots, r_m$:*

$$(5.3) \qquad \pi_1(X, x_0) \cong \langle x_1, x_2, \cdots, x_k \rangle / N(r_1, r_2, \cdots, r_m).$$

PROOF. According to van Kampen's theorem (ii), the fundamental group $\pi_1(X^2, x_0)$ of the 2-skeleton X^2 is certainly given by the right-hand side of the above isomorphism. The whole X is obtained by attaching cells of dimensions greater than or equal to 3 on X^2, so $\pi_1(X, x_0)$ is isomorphic to $\pi_1(X^2, x_0)$ by van Kampen's theorem (i). This completes the proof. $\square$

The right-hand side of the isomorphism (5.3) is regarded as a group obtained from the free group $\langle x_1, x_2, \cdots, x_k \rangle$ by adding the relations $r_1 = 1, r_2 = 1, \cdots, r_m = 1$, and it is sometimes denoted by

$$\langle\ x_1, x_2, \cdots, x_k\ |\ r_1 = r_2 = \cdots = r_m = 1\ \rangle.$$

This is called a *presentation* of a group by generators and relators.

COROLLARY 5.8. *The group $\pi_1(X, x_0)^{\mathrm{ab}}$ obtained from the fundamental group $\pi_1(X, x_0)$ by adding relations declaring that "all elements commute with each other" is isomorphic to the 1-dimensional homology group $H_1(X)$.*

The group $\pi_1(X, x_0)^{\mathrm{ab}}$ is called the *abelianization* of $\pi_1(X, x_0)$. The notation "ab" in the superscript is an abbreviation of Abel.

PROOF. It can be assumed that the 0-skeleton consists only of a single point x_0 by deforming X without altering its homotopy type. Then, the abelianization of $\langle x_1, x_2, \cdots, x_k \rangle$ that appears in the right-hand side of the isomorphism (5.3) can be identified with the 1-dimensional cycle group $Z_1(X)$. With this identification, r_i can be identified with the boundary $\partial_2(\langle e_i^2 \rangle)$, where the 2-cell e_i^2 is the one with boundary r_i. From this we see that $\pi_1(X, x_0)^{\mathrm{ab}} \cong H_1(X)$. $\square$

COROLLARY 5.9. *A connected manifold M with a handle decomposition without 1-handles is simply connected.*

PROOF. According to Theorem 4.18 in Chapter 4, such an M is homotopy equivalent to a cell complex X without 1-cells. Then the 1-skeleton X^1 is a single point. By Theorem 5.7, $\pi_1(X, x_0)$ is a quotient group of $\pi_1(X^1, x_0)$, so in this case it is the trivial group. □

EXAMPLE 5.10. The complex projective spaces $\mathbb{C}P^n$ and the special unitary groups $SU(n)$ are simply connected.

In fact, we showed that $\mathbb{C}P^n$ and $SU(n)$ have handle decompositions without 1-handles, in Chapter 3, Section 2.

5.2. Closed surfaces and 3-dimensional manifolds

In this section, we first prove the classification theorem for closed surfaces, and then describe Heegaard decompositions of 3-manifolds.

(a) Closed surfaces

A connected, compact, closed 2-manifold is called a closed surface.

THEOREM 5.11 (Classification of closed surfaces). *Any closed surface M is diffeomorphic to one and only one of the following.*
(i) *The float for g people, Σ_g, $g = 0, 1, 2, \cdots$.*
(ii) *The connected sum $P^2 \# P^2 \# \cdots \# P^2$ of k copies of the projective plane P^2, $k = 1, 2, \cdots$.*
Here, the family (i) *consists of orientable closed surfaces, and the family* (ii) *consists of non-orientable closed surfaces.*

The float for g people Σ_g means the orientable closed surface of genus g, depicted in Figure 1.6 in Chapter 1 (for $g = 2, 3$).

In particular, Σ_0 is the sphere S^2, and Σ_1 is the torus T^2.

The *connected sum* of two closed surfaces M_1 and M_2,

$$M_1 \# M_2,$$

is a closed surface obtained from M_1 and M_2 as follows: remove the interior of smoothly embedded 2-disks $D_1^2 \subset M_1$ and $D_2^2 \subset M_2$ from M_1 and M_2 respectively, to get surfaces with boundary

$$M_1 - \mathrm{int} D_1^2 \quad \text{and} \quad M_2 - \mathrm{int} D_2^2,$$

and then glue these along their boundary circles S^1 (see Figure 5.2). If M_1 and M_2 are oriented, a diffeomorphism

$$\partial(M_1 - \mathrm{int} D_1^2) \to \partial(M_2 - \mathrm{int} D_2^2),$$

used when the boundaries are glued, must be chosen in such a way that the resulting connected sum $M_1 \# M_2$ inherits the same orientations as M_1 and M_2.

The definition of the connected sum for more than two closed surfaces $M_1 \# M_2 \# \cdots \# M_n$ is similar.

For $g \geq 1$, Σ_g is diffeomorphic to the connected sum of g tori $T^2 \# T^2 \# \cdots \# T^2$ (g copies).

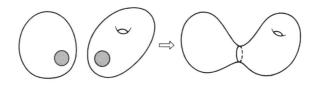

FIGURE 5.2. Connected sum $M_1 \# M_2$

As we will see in the following proof, the Euler number of Σ_g is $2 - 2g$, and the Euler number of the connected sum $P^2 \# P^2 \# \cdots \# P^2$ (k copies) is $2 - k$. Hence as a corollary of Theorem 5.11, we obtain

COROLLARY 5.12. *All closed surfaces are classified by orientability and the Euler number.*

Now we prove Theorem 5.11.

PROOF. By Theorem 2.20 in Chapter 2, there exists a Morse function $f : M \to \mathbb{R}$ on a closed surface M. By choosing an appropriate f, we can assume that there is only one 0-handle and one 2-handle in the handle decomposition of f (Theorem 3.35 in Chapter 3). Therefore, M has a handle decomposition of the following form:

$$(5.4) \qquad M = h^0 \cup (h_1^1 \sqcup h_2^1 \sqcup \cdots \sqcup h_k^1) \sqcup h^2.$$

Note that the Euler number $\chi(M)$ of M is written as

$$(5.5) \qquad \chi(M) = 2 - k,$$

using the number k of 1-handles, from Corollary 4.19 in Chapter 4.

Let us start our investigation from small values of k.

If $k = 0$, then M is diffeomorphic to the 2-sphere S^2, $M \cong S^2$, by Theorem 1.16. In this case, $\chi(M) = 2$.

If $k = 1$, M is diffeomorphic to the projective plane P^2, $M \cong P^2$, by Example 3.9 in Chapter 3. In this case, $\chi(M) = 1$. The projective plane P^2 contains a Möbius band, so it is non-orientable.

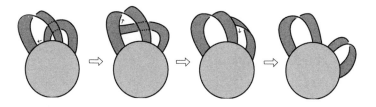

FIGURE 5.3. Uncrossing feet of handles

Let us compute the fundamental group $\pi_1(P^2, x_0)$ using van Kampen's theorem. Let N be a Möbius band, and regard $P^2 = N \cup D^2$. Take the base point x_0 on the boundary ∂N. Since N is homotopy equivalent to the central circle S^1, $\pi_1(N, x_0)$ is isomorphic to $\mathbb{Z}$. Then D^2 is attached along the boundary of N. By looking at how it is actually attached, we see that the boundary of N wraps around the circle that is the center line of N twice (up to homotopy), and we see that twice the generator of $\pi_1(N, x_0)$ is the identity element (in this case 0), by van Kampen's theorem. Hence we obtain

$$\pi_1(P^2, x_0) \cong \mathbb{Z}_2.$$

This completes our investigation for the case $k = 1$.

There are two cases to be considered for $k = 2$: the case where 1-handles are attached to the 0-handle in an orientation-preserving way, and the case of an orientation-reversing way as for P^2.

Let us consider the case where at least one (say, h_1) among two 1-handles is attached in an orientation-reversing way. If a foot of the other handle h_2 is crossed with a foot of h_1 and attached, then the foot of h_1 and the foot of h_2 can be uncrossed by sliding the foot of h_2 along h_1 (Figure 5.3).

Suppose that the feet of two handles h_1, h_2 are not crossed. Since h_1 is attached in an orientation-reversing way, h_2 must be attached in an orientation-reversing way as well. Otherwise the boundary $\partial(h^0 \cup h_1 \cup h_2)$ becomes disconnected, so that we will not be able to cap it off with a single 2-handle to construct a closed surface. After all, we see that h_1 and h_2 must be attached as shown in Figure 5.4.

We see that we obtain $P^2 \# P^2$ by capping off the surface in Figure 5.4 with a 2-handle, so that the closed surface in this case is diffeomorphic to $P^2 \# P^2$, $M \cong P^2 \# P^2$. The Euler number is $\chi(M) = 0$. The surface $P^2 \# P^2$ is sometimes called a *Klein bottle*.

FIGURE 5.4. We obtain $P^2 \# P^2$ by capping this off

Let us compute the fundamental group $\pi_1(P^2 \# P^2, x_0)$ by taking the base point x_0 at the center of the 0-handle h^0 in Figure 5.4.

Since the handlebody in Figure 5.4 is homotopy equivalent to the bouquet of two circles, its fundamental group is the free group of rank 2, $\langle x_1, x_2 \rangle$. The generators x_1, x_2 correspond to the cores of 1-handles h_1, h_2, respectively. When a 2-handle is attached along the boundary of Figure 5.4 to cap it off, the relation (the loop corresponding to this boundary)$= 1$ is introduced. By actually tracking the boundary, we see that the loop corresponding to the boundary is conjugate to $x_1^2 x_2^2$. Hence we obtain

$$\pi_1(P^2 \# P^2, x_0) \cong \langle \, x_1, x_2 \mid x_1^2 x_2^2 = 1 \, \rangle.$$

By Corollary 5.8, the abelianization of the fundamental group is the first homology group, so we find that

$$H_1(P^2 \# P^2) \cong \mathbb{Z} \oplus \mathbb{Z}_2.$$

The other case we have to consider when $k = 2$ is the case where both h_1 and h_2 are attached in an orientation-preserving way. In this case, a foot of h_1 and a foot of h_2 must be crossed as shown in Figure 5.5, since the boundary $\partial(h^0 \cup h_1 \cup h_2)$ is disconnected otherwise.

We obtain a torus T^2 by capping off Figure 5.5 along its boundary with a 2-handle. In fact, it can be confirmed that a torus T^2 with the interior $\text{int}(D^2)$ of a 2-disk removed, $T^2 - \text{int}(D^2)$, is diffeomorphic to the handlebody of Figure 5.5, by deformations described in Figure 5.6.

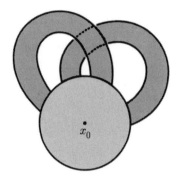

FIGURE 5.5. We obtain a torus T^2 by capping this off

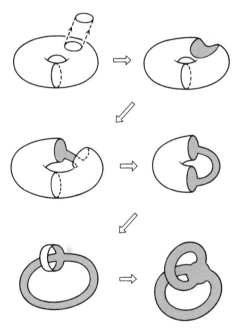

FIGURE 5.6. The surface obtained from a torus by removing the interior of a disk

As before, let us compute the fundamental group $\pi_1(T^2, x_0)$ of the torus T^2 by choosing the base point x_0 at the center of the 0-handle h^0 as in Figure 5.5. Since the handlebody in Figure 5.5 is also homotopy equivalent to the bouquet of two circles, its fundamental group is again the free group of rank 2, $\langle x_1, x_2 \rangle$. Furthermore, the loop corresponding to the boundary of this handlebody is conjugate to $x_1 x_2 x_1^{-1} x_2^{-1}$, so that we obtain

$$\pi_1(T^2, x_0) \cong \langle\ x_1, x_2 \mid x_1 x_2 x_1^{-1} x_2^{-1} = 1\ \rangle.$$

The term $x_1 x_2 x_1^{-1} x_2^{-1}$ which appeared here is denoted by $[x_1, x_2]$, and is called the *commutator* of x_1 and x_2. The relation that a commutator is equal to the identity element ($[x_1, x_2] = 1$) is equivalent to the relation $x_1 x_2 = x_2 x_1$ (x_1 and x_2 commute), so we see that the fundamental group $\pi_1(T^2, x_0)$ is isomorphic to the free abelian group of rank 2, $\mathbb{Z} \oplus \mathbb{Z}$.

Since the fundamental group of the torus is an abelian group, its homology group $H_1(T^2)$ is isomorphic to the fundamental group:

$$H_1(T^2) \cong \mathbb{Z} \oplus \mathbb{Z}.$$

In summary, there are two cases for $k = 2$, the orientable case gives us the torus T^2, and the non-orientable case gives us the Klein bottle $P^2 \# P^2$. Both have Euler number 0, but a torus and a Klein bottle are not homeomorphic since their homology groups are not isomorphic.

In the case $k = 3$, the boundary $\partial(h^0 \cup h_1 \cup h_2 \cup h_3)$ is disconnected if all three 1-handles h_1, h_2, h_3 are attached in an orientation-preserving way (Exercise 5.1). Therefore, at least one 1-handle (say, h_1) is attached in an orientation-reversing way. Then, in a similar way to the case $k = 2$, the feet of h_2 and h_3 can be slid off of the feet of h_1 so that all feet are uncrossed, and we obtain Figure 5.7 at the end. By capping this off with a 2-handle, we obtain $P^2 \# P^2 \# P^2$ (Figure 5.7).

We compute that the homology group is

$$H_1(P^2 \# P^2 \# P^2) \cong \mathbb{Z} \oplus \mathbb{Z} \oplus \mathbb{Z}_2$$

by abelianizing the fundamental group. The Euler characterictic is $\chi(M) = -1$.

In fact, there is another handlebody with three 1-handles, as depicted in Figure 5.8, which has a connected boundary $\partial(h^0 \cup h_1 \cup h_2 \cup h_3)$. It can be seen, however, that the handlebody of Figure 5.8 is diffeomorphic to the handlebody of Figure 5.7 by sliding feet

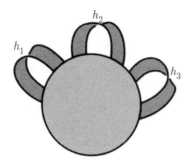

FIGURE 5.7. We obtain $P^2 \# P^2 \# P^2$ by capping this off

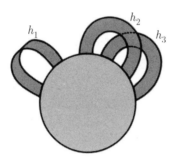

FIGURE 5.8. We obtain $P^2 \# T^2$ by capping this off

of 1-handles (Exercise 5.2). We obtain $P^2 \# T^2$ by capping off the handlebody of Figure 5.8 with a 2-handle, so from the above remark, we see that

$$P^2 \# T^2 \cong P^2 \# P^2 \# P^2.$$

In summary, we have $M \cong P^2 \# P^2 \# P^2$ in the case $k = 3$, and the Euler number is $\chi(M) = -1$. In this case M is non-orientable.

We find the following fact in the case $k \geq 4$ as well, by sliding handles (strictly speaking, using mathematical induction): if there is a 1-handle in the handlebody (5.4) which is attached in an orientation-reversing way, then $h^0 \cup h_1 \cup h_2 \cup \cdots \cup h_k$ is diffeomorphic to the handlebody of Figure 5.9. In this case M is non-orientable, and

$$M \cong P^2 \# P^2 \# \cdots \# P^2 \quad (k \text{ copies of } P^2),$$

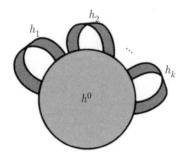

FIGURE 5.9. We obtain $P^2 \# P^2 \# \cdots \# P^2$ (k copies of P^2) by capping this off

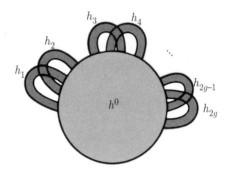

FIGURE 5.10. We obtain $T^2 \# T^2 \# \cdots \# T^2$ (g copies of T^2) by capping this off

and the Euler number is $\chi(M) = 2 - k$. The homology group is isomorphic to the direct sum of $k - 1$ copies of $\mathbb{Z}$ and a single copy of $\mathbb{Z}_2$.

If there is no orientation-reversing 1-handle, then $h^0 \cup h_1 \cup h_2 \cup \cdots \cup h_k$ is diffeomorphic to the handlebody depicted in Figure 5.10. In this case, k is even (say, $2g$), M is orientable, and

$$M \cong T^2 \# T^2 \# \cdots \# T^2 \quad (g \text{ copies of } T^2).$$

This is a float for g people. The Euler number is $\chi(M) = 2 - 2g$. The homology group is isomorphic to the direct sum of $2g$ copies of $\mathbb{Z}$.

An orientable closed surface and a non-orientable closed surface cannot be homeomorphic, since the homology groups of orientable closed surfaces do not contain torsion subgroups and those of non-orientable closed surfaces contain a torsion subgroup ($\cong \mathbb{Z}_2$). This completes the proof of Theorem 5.11. □

(b) 3-dimensional manifolds

Let M be a connected, orientable, closed 3-manifold. Let $f : M \to \mathbb{R}$ be a Morse function such that there are only one 0-handle and one 3-handle in the handlebody decomposition associated with f:

$$(5.6) \qquad M = h^0 \cup (h_1^1 \sqcup \cdots h_{k_1}^1) \cup (h_1^2 \sqcup \cdots h_{k_2}^2) \cup h^3.$$

If we express the Euler number of M using the numbers k_1 and k_2 of 1-handles and 2-handles respectively, then we obtain

$$\chi(M) = 1 - k_1 + k_2 - 1 = -k_1 + k_2$$

(by Corollary 4.19 in Chapter 4).

On the other hand, the Euler number of an odd-dimensional manifold is always 0 (Exercise 4.3), and we obtain $k_1 = k_2$. Let this value be k.

Let N be the subhandlebody of the handle decomposition (5.6) consisting of a 0-handle and 1-handles. By the orientability of N, N has the form depicted in Figure 5.11. In the terminology specifically used in 3-manifold theory, a 3-dimensional handlebody as in Figure 5.11 is called a *handlebody* of genus g. This name is confusing compared with more general handlebodies discussed in Chapter 3, but in 3-manifold theory, handlebodies always mean the ones depicted in Figure 5.11. This is sometimes denoted by H_k. Its boundary ∂H_k is the orientable closed surface Σ_k of genus k.

Let $N^* = M - \mathrm{int}(N)$; then N^* is the union of a 0-handle and 1-handles in the handle decomposition associated with $-f$. The 1-handles of N^* corresponds to the 2-handles of the original handle decompostion (5.6), so that the genus of N^* is equal to the number of 2-handles of N, which is the genus, k, of N. Therefore N^* is diffeomorphic to N. Thus, M can be described as the union

$$(5.7) \qquad M = N \cup_\varphi N^*$$

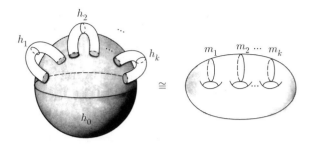

FIGURE 5.11. A handlebody of genus k and its meridians

obtained from the handlebodies N and N^* of the same genus k by gluing along their boundaries by a diffeomorphism

$$\varphi : \partial N^* \to \partial N.$$

This description of M is called a *Heegaard splitting of genus k*.

The belt spheres of 1-handles contained in the handlebody N of genus k form k copies of circles in the boundary ∂N. Such a circle is called a *meridian*. They are denoted by

$$m_1, \ m_2, \ \cdots, \ m_k$$

(see Figure 5.11). Similarly, meridians $m_1^*, m_2^*, \cdots, m_k^*$ are defined for N^*. Since the 1-handles of N^* are 2-handles in the handle decomposition (5.6) of M, the meridians of N^* are attaching spheres in the handle decomposition (5.6).

In the Heegaard splitting (5.7), we look at the images in ∂N of the meridians $m_1^*, m_2^*, \cdots, m_k^*$ under the attaching map $\varphi : \partial N^* \to \partial N$. Denote the images by

$$\mu_1 = \varphi(m_1^*), \quad \mu_2 = \varphi(m_2^*), \quad \cdots, \quad \mu_k = \varphi(m_k^*).$$

DEFINITION 5.13. The handlebody N of genus k, together with the closed curves $\mu_1, \mu_2, \cdots, \mu_k$ on the boundary ∂N obtained as above,

$$(N; \mu_1, \mu_2, \cdots, \mu_k),$$

is called a *Heegaard diagram of M*.

Since handle decompositions of M are not unique, Heegaard diagrams of M are not unique either. For the opposite correspondence, however, the following can be stated.

THEOREM 5.14. *A Heegaard diagram* $(N; \mu_1, \mu_2, \cdots, \mu_k)$ *determines the diffeomorphism type of the original manifold M uniquely. Furthermore, for an isotopy $\{h_t\}_{t \in J}$ of ∂N with $h_0 = \mathrm{id}_{\partial N}$, consider the Heegaard diagram $(N; h_1(\mu_1), h_1(\mu_2), \cdots, h_1(\mu_k))$ determined by the images of $(\mu_1, \mu_2, \cdots, \mu_k)$ under the diffeomorphism $h_1 : \partial N \to \partial N$ which corresponds to $t = 1$ of the isotopy. Then the 3-manifold M' determined by this Heegaard diagram is also diffeomorphic to M.*

PROOF. First, we prove that the diffeomorphism type of the handlebody

$$N(\mu_1, \mu_2, \cdots, \mu_k) = N \cup (h_1^2 \sqcup h_2^2 \sqcup \cdots \sqcup h_k^2),$$

obtained from N by attaching 2-handles $h_1^2, h_2^2, \cdots, h_k^2$ along $\mu_1, \mu_2, \cdots, \mu_k$, is determined only by $\mu_1, \mu_2, \cdots, \mu_k$. In fact, to attach a 2-handle $h_i^2 = D^2 \times D^1$ along μ_i, an attaching map $\varphi_i : \partial D^2 \times D^1 \to \partial N$ must be specified in such a way that $\varphi_i(\partial D^2 \times \mathbf{0}) = \mu_i$. However, using the "theorem on uniqueness of tubular neighborhood" (see [7], p. 110) and the fact that the co-core D^1 of the handle is of dimension 1 in our case, we can prove that such a φ_i is unique up to isotopy of ∂N. (To be precise, there remains the freedom of composing the map

$$r_1 \times \mathrm{id}$$

which is the product of an orientation-reversing map $r_1 : \partial D^2 \to \partial D^2$ of a circle ∂D^2 and $\mathrm{id} : D^1 \to D^1$, or the map

$$\mathrm{id} \times r_2$$

which is the product of a map $r_2 : D^1 \to D^1$ of interchanging the two endpoints of D^1 and $\mathrm{id} : \partial D^2 \to \partial D^2$. However, these maps $r_1 \times \mathrm{id}$ and $\mathrm{id} \times r_2$ extend to difeomorphisms from a 2-handle $D^2 \times D^1$ to itself, so that the diffeomorphism type of the resulting handlebody is not affected by composing these maps with the attaching map φ_i.) Furthermore, the diffeomorphism type of the resulting $N \cup (h_1^2 \sqcup h_2^2 \sqcup \cdots \sqcup h_k^2)$ is not altered by changing the handle attaching map φ_i by isotopy (Theorem 3.30). This proves the above.

To obtain M, we attach D^3 to $N(\mu_1, \mu_2, \cdots, \mu_k)$ with a diffeomorphism

$$\psi : \partial D^3 \to \partial N(\mu_1, \mu_2, \cdots, \mu_k).$$

For a different choice of diffeomorphism ψ', the composition $\psi^{-1} \circ \psi' : \partial D^3 \to \partial D^3$ extends to a diffeomorphism $D^3 \to D^3$ ([22]), so that the diffeomorphism type of M does not depend on the choice of ψ either. This proves the first half of Theorem 5.14.

The second half can be proved similarly to Theorem 3.21.   □

REMARK. When a handlebody N of genus k and k mutually dis-joint simple closed curves $\mu_1, \mu_2, \cdots, \mu_k$ on ∂N are given arbitrarily, $(N; \mu_1, \mu_2, \cdots, \mu_k)$ does not necessarily represent a Heegaard diagram of some 3-manifold M. A necessary and sufficient condition for this to be a Heegaard diagram is that the boundary $\partial N(\mu_1, \mu_2, \cdots, \mu_k)$ of the handlebody $N(\mu_1, \mu_2, \cdots, \mu_k)$, obtained by attaching k copies of 2-handles on N along $\mu_1, \mu_2, \cdots, \mu_k$ is the 2-sphere S^2.

Let $(N; \mu_1, \mu_2, \cdots, \mu_k)$ be a Heegaard diagram of genus k of a closed 3-manifold M. Take the base point x_0 on ∂N and con-sider the fundamental group $\pi_1(N, x_0)$. Then this is the free group $\langle x_1, x_2, \cdots, x_k \rangle$ of rank k. The fundamental group $\pi_1(M, x_0)$ of M is the free group $\langle x_1, x_2, \cdots, x_k \rangle$ with relations added that declare that all the elements of the free group corresponding to the loops $\mu_1, \mu_2, \cdots, \mu_k$ are equal to 1. The element of the free group corre-sponding to each loop μ_i can be obtained by reading off the intersec-tion between μ_i and meridians $m_1, m_2, \cdots, m_k$ of N.

EXAMPLE 5.15 (Projective space P^3 of dimension 3). We seek a Heegaard diagram of the 3-dimensional projective space P^3. Accord-ing to Section 2 of Chapter 3, P^3 has a handle decomposition of the form

$$P^3 = h^0 \cup h^1 \cup h^3 \cup h^3.$$

Therefore, P^3 should have a Heegaard splitting of genus 1.

A handlebody N of genus 1 is diffeomorphic to $S^1 \times D^2$. This is called a *solid torus*. When we fix a single point θ_0 on S^1, $\{\theta_0\} \times \partial D^2$ is a meridian. A simple closed curve l on ∂N, which intersects m in a single point transversely, is called a *longitude*. Although the choice of a meridian m is unique up to isotopy of the surface of a solid torus, the choice of a longitude is not. Here, we fix a point ϕ_0 on ∂D^2, and take a closed curve of the form $S^1 \times \{\phi_0\}$ as our longitude l.

The fundamental group $\pi_1(N, x_0)$ with the base point $x_0 = (\theta_0, \phi_0)$ on ∂N is computed as $\pi_1(N, x_0) \cong \mathbb{Z}$. A longitude is a generator. The 3-dimensional projective space P^3 has a cell decomposition of the form

$$P^3 = e^0 \cup e^1 \cup e^2 \cup e^3,$$

so that the union of the first three cells $e^0 \cup e^1 \cup e^2$ gives a cell de-composition of the 2-dimensional projective space P^2 (Exercise 5.3).

Therefore, by van Kampen's Theorem 5.6, we obtain

$$\pi_1(P^3, x_0) \cong \pi_1(P^2, x_0) \cong \mathbb{Z}_2,$$

since P^3 is P^2 with a 3-cell attached.

From this fact, a simple closed curve μ that gives a Heegaard diagram of P^3 must intersect the meridian m twice transversely, in the same direction. Such a loop can be matched to the following loop $f : (I, \partial I) \to (S^1 \times \partial D^2, x_0)$,

$$f(t) = (4\pi t, \pm 2\pi t),$$

after isotopy of ∂N and using a diffeomorphism $f_n | S^1 \times \partial D^2 : S^1 \times \partial D^2 \to S^1 \times \partial D^2$ of Section 5.3 (a). In the above definition of f, the coordinates for $S^1 \times \partial D^2$ are given by angles, and the base point was taken at $x_0 = (0, 0)$. It is not important whether we take $+$ or $-$ in the second coordinate, since it depends on how we choose orientations of $S^1 \times D^2$ and P^3.

Thus a Heegaard diagram of P^3 as shown in Figure 5.12 has been obtained.

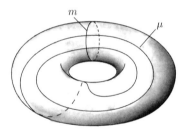

FIGURE 5.12. A Heegaard diagram of the projective space P^3

EXAMPLE 5.16 (Lens spaces). Identify the 3-sphere S^3 with the unit sphere $\{(z_1, z_2) \,|\, |z_1|^2 + |z_2|^2 = 1\}$ of $\mathbb{C}^2$. Let p be a natural number greater than or equal to 2, and q a natural number relatively prime to p. For simplicity, assume $1 \le q \le p - 1$.

Define two points (z_1, z_2) and (z'_1, z'_2) of S^3 to be equivalent, $(z_1, z_2) \sim (z'_1, z'_2)$, when

$$(z'_1, z'_2) = (\omega^k z_1, \omega^{qk} z_2)$$

for some integer k, where $\omega = \exp(2\pi\sqrt{-1}/p)$. This relation $\sim$ certainly is an equivalence relation, and the quotient space of S^3 by this equivalence relation is called the *lens space* of type (p, q), and denoted by

$$L(p, q) = S^3 / \sim .$$

A lens space is an orientable closed 3-manifold. The sphere S^3 is a p-fold cover of $L(p, q)$, and $\pi_1(L(p, q), x_0) \cong \mathbb{Z}_p$.

The projective space P^3 can be regarded as the lens space $L(2, 1)$ of type $(2, 1)$.

We describe, without a proof, a Heegaard diagram of the lens space $L(p, q)$.

Lens spaces can be described by Heegaard diagrams of genus 1. Let $N = S^1 \times D^2$. In the same way as in the projective space case, give the coordinates (θ, ϕ) of pairs of angles on the boundary $\partial N = S^1 \times \partial D^2$, and define a loop $f : (I, \partial I) \to (S^1 \times \partial D^2, x_0)$ by

$$f(t) = (2\pi p t, 2\pi q t),$$

where $x_0 = (0, 0)$. Let μ be the simple closed curve represented by this loop. Then $(N; \mu)$ is a Heegaard diagram of $L(p, q)$.

5.3. 4-dimensional manifolds

(a) Heegaard diagrams for 4-dimensional manifolds

Let M be a connected, orientable, closed 4-manifold. We can choose a nice Morse function $f : M \to \mathbb{R}$ so that the associated handlebody has a single 0-handle and a single 4-handle (Theorem 3.35). In this case, the handle decomposition can be written as

$$
(5.8) \quad
\begin{aligned}
M = h^0 \ &\cup (h_1^1 \sqcup h_2^1 \sqcup \cdots \sqcup h_{k_1}^1) \\
&\cup (h_1^2 \sqcup h_2^2 \sqcup \cdots \sqcup h_{k_2}^2) \cup (h_1^3 \sqcup h_2^3 \sqcup \cdots \sqcup h_{k_3}^3) \cup h^4.
\end{aligned}
$$

In the handle decomposition (5.8), let N be the union of the 0-handle and 1-handles. In analogy with handlebodies in 3-manifold theory, we call this N a 4-dimensional handlebody. Of course, this definition is more restrictive than that of the handlebodies discussed in Chapter 3.

If the number k_1 of 1-handles is equal to 1, then N is a single 4-disk D^4 with a single 1-handle $D^1 \times D^3$ attached in such a way that the result is orientable. In this case, we see that $N \cong S^1 \times D^3$, which is the 4-dimensional version of a solid torus. The boundary ∂N is diffeomorphic to $S^1 \times S^2$.

More generally, for $k_1 \geq 1$, we have the following.

LEMMA 5.17. *The boundary ∂N is diffeomorphic to the connected sum of k_1 copies of $S^1 \times S^2$:*

$$\partial N \cong (S^1 \times S^2)\#(S^1 \times S^2)\# \cdots \#(S^1 \times S^2) \qquad (k_1 \text{ copies of } S^1 \times S^2).$$

The conncted sum was defined for closed surfaces in the preceding section, and we define it here for manifolds of general dimension.

DEFINITION 5.18 (Connected sum). Let M_1 and M_2 be oriented m-manifolds. Let $f_1 : D^m \to M_1$ be a smooth, orientation-preserving embedding, and $f_2 : D^m \to M_2$ a smooth, orientation-reversing embedding. We remove the interior of the embedded disks and identify the boundaries of $M_1 - f_1(\text{int } D^m)$ and $M_2 - f_2(\text{int } D^m)$ by a diffeomorphism

$$\psi = f_2 \circ f_1^{-1}|f_1(\partial D^m) : f_1(\partial D^m) \to f_2(\partial D^m).$$

The resulting manifold

$$(M_1 - f_1(\text{int } D^m)) \cup_\psi (M_2 - f_2(\text{int } D^m))$$

is called the *connected sum*, and is denoted by

$$M_1 \# M_2.$$

The connected sum $M_1 \# M_2$ receives the orientation which extends both orientations of M_1 and M_2 simultaneously.

The diffeomorphism type of the connected sum $M_1 \# M_2$ with orientation considered is determined by those of M_1 and M_2, and does not depend on the embeddings f_1 and f_2 used for construction (see [18]).

Let us go back to the handle decomposition (5.8). The subhandlebody $N^{(2)}$ with the 0-handle, the 1-handles, and 2-handles all attached is obtained from N by attaching all the 2-handles:

$$N^{(2)} = N \cup (h_1^2 \sqcup h_2^2 \sqcup \cdots \sqcup h_{k_2}^2),$$

where the attaching map of the i-th 2-handle $h_i^2 = D^2 \times D^2$ is a smooth embedding

$$\varphi_i : \partial D^2 \times D^2 \to \partial N.$$

Recall that in the case of 3-dimensional Heegaard diagrams, an attaching map $\varphi : \partial D^2 \times D^1 \to \partial N$ of a 2-handle was determined up to isotopy by the image $\varphi(\partial D^2 \times \mathbf{0})$ of the attaching sphere (circle) $\partial D^2 \times \mathbf{0}$. This is due to the fact that there is a unique diffeomorphism, up to isotopy, from the annulus $\partial D^2 \times D^1$ to itself, which preserves the orientation and the direction of the circle ∂D^2.

In the case of dimension 4, we have $\dim(\partial N) = 3$, and moreover $\varphi(\partial D^2 \times D^2)$ embedded in ∂N is a 3-dimensional solid torus ($\cong S^1 \times D^2$). There are infinitely many varieties, even up to isotopy, of diffeomorphisms of a solid torus to itself which preserve the orientation, so that the attaching map $\varphi_i : \partial D^2 \times D^2 \to \partial N$ of a 2-handle is not determined only by specifying the image $\varphi_i(\partial D^2 \times \mathbf{0})$ of the attaching sphere (circle) $\partial D^2 \times \mathbf{0}$.

Let us observe that there are infinitely many orientation-preserving diffeomorphisms from the solid torus $S^1 \times D^2$ to itself.

Take the coordinate of angles θ for S^1, and the polar coordinates (r, ϕ) for the disk D^2, where r is the distance from the origin $\mathbf{0}$, and assume that $0 \leq r \leq 1$. Let n be a natural number, and define a diffeomorphism $f_n : S^1 \times D^2 \to S^1 \times D^2$ by the expression

$$f_n(\theta, (r, \phi)) = (\theta, (r, \phi + n\theta)).$$

This diffeomorphism rotates n times the disk D^2 which transversely intersects S^1, while going around S^1 once.

It is known that f_n and f_m are not related by isotopy if n and m are distinct. It is also known that any diffeomorphism from $S^1 \times D^2$ to itself, which is orientation-preserving and also preserves the orientation of the circle S^1, can be matched to one of the f_n's by isotopy.

By taking a composition of an attaching map $\varphi_i : \partial D^2 \times D^2 \to \partial N$ with f_n and taking the map $\varphi_i \circ f_n : \partial D^2 \times D^2 \to \partial N$ as a new attaching map, we obtain a map with the same image of the core sphere (circle):

$$\varphi_i(\partial D^2 \times \mathbf{0}) = \varphi_i \circ f_n(\partial D^2 \times \mathbf{0}).$$

As the diffeomorphism types of the resulting manifolds may vary depending on whether the attaching map of a 2-handle is φ_i or $\varphi_i \circ f_n$, we have to eliminate the ambiguity of composing f_n to determine the resulting manifold uniquely. A method for this purpose is to specify a "frame" along the core circle $\varphi_i(\partial D^2 \times \mathbf{0})$.

Regard D^2 as the unit disk $\{(x, y)|x^2 + y^2 \leq 1\}$ in the xy-plane. Fix two unit vectors $\mathbf{e}_1$ and $\mathbf{e}_2$ in D^2, that start from the origin $\mathbf{0}$ and are normal to each other. The direct products $\partial D^2 \times \mathbf{e}_1$ and $\partial D^2 \times \mathbf{e}_2$ can be regarded as the standard two vector fields $\tilde{\mathbf{e}}_1, \tilde{\mathbf{e}}_2$ in $\partial D^2 \times D^2$ along the core circle $\partial D^2 \times \mathbf{0}$.

These vector fields are perpendicular to the core circle, and perpendicular to each other.

DEFINITION 5.19 (Frame). Suppose that there are two vector fields $\mathbf{v}_1$ and $\mathbf{v}_2$ along a smooth simple closed curve C in a 3-manifold K, such that they are never tangent to C, and linearly independent everywhere. Then $\{\mathbf{v}_1, \mathbf{v}_2\}$ is called a *frame* of C. Two frames $\{\mathbf{v}_1, \mathbf{v}_2\}$ and $\{\mathbf{w}_1, \mathbf{w}_2\}$ are said to be *equivalent* if they are related by an isotopy fixing C. A closed curve C with an equivalence class of frames specified is called a *framed closed curve* (see Figure 5.13).

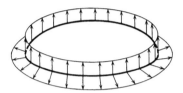

FIGURE 5.13. A framed closed curve

When an attaching map $\varphi_i : \partial D^2 \times D^2 \to \partial N$ is given, by mapping the standard frame $\{\tilde{\mathbf{e}}_1, \tilde{\mathbf{e}}_2\}$ in $\partial D^2 \times D^2$ by φ_i, we obtain a frame $\{\mathbf{v}_1, \mathbf{v}_2\}$ $(= \{\varphi_i(\tilde{\mathbf{e}}_1), \varphi_i(\tilde{\mathbf{e}}_2)\})$ of the simple closed curve $\varphi_i(\partial D^2 \times \mathbf{0})$. It is sufficient to specify the core circle $\varphi_i(\partial D^2 \times \mathbf{0})$ and the equivalence class of a frame $\{\mathbf{v}_1, \mathbf{v}_2\}$ to specify the isotopy class of an attaching map φ_i.

This way, for each $i = 1, 2, \cdots, k_2$, the isotopy class of the attaching map φ_i is specifed by regarding the simple closed curve $\varphi_i(\partial D^2 \times \mathbf{0})$ in the boundary ∂N as a framed closed curve, so that the diffeomorphism type of the subhandlebody $N^{(2)}$, with all 2-handles attached, is determined.

Furthermore, we have the following result ([16]).

LEMMA 5.20. *In the handle decomposition (5.8), the diffeomorphism type of the connected, orientable, closed 4-manifold M is determined by the subhandlebody $N^{(2)}$, with all 0- through 2-handles attached.*

We sketch the proof. The complement $N^* = M - \text{int}(N^{(2)})$ is a 4-dimensional handlebody with a 0-handle and 1-handles in the handle decomposition associated with the "upside-down" Morse function $-f$. The diffeomorphism type of N^* is determined by the number of 1-handles associated with $-f$, so that it is also determined by $\partial N^* = \partial N^{(2)}$ (cf. Lemma 5.17). Furthermore, any diffeomorphism

$h : \partial N^* \to \partial N^*$, from the boundary ∂N^* of a 4-dimensional handle-body N^* to itself, extends to a diffeomorphism $\tilde{h} : N^* \to N^*$, from N^* to itself, by the Laudenbach-Poenaru theorem [8]. Therefore, when we construct $N'^{(2)}$ from a handle decomposition of another closed 4-manifold M', if there is a diffeomorphism $h : N^{(2)} \to N'^{(2)}$, then h extends to a diffeomorphism $\tilde{h} : M \to M'$. This proves Lemma 5.20.

As we explained right before Lemma 5.20, the diffeomorphism type of $N^{(2)}$ is determined by the 4-dimensional handlebody N and *framed* closed curves $C_1, C_2, \cdots, C_{k_2}$ in the boundary ∂N, where the notation is abbreviated by $C_i = \varphi_i(\partial D^2 \times \mathbf{0})$. Considering Lemma 5.20, the following definition would be natural (see Montesinos [16]).

DEFINITION 5.21. $(N; C_1, C_2, \cdots, C_{k_2})$ is called a *Heegaard dia-gram* of a 4-manifold M.

The diffeomorphism type of an orientable closed 4-manifold is uniquely determined by a Heegaard diagram.

(b) The case $N = D^4$

In a handle decomposition (5.8) of a closed 4-manifold M, con-sider the special case where there is no 1-handle ($k_1 = 0$). In this case $N = D^4$, so that a Heegaard diagram of M has the form

$$(D^4; C_1, C_2, \cdots, C_k),$$

where $C_1, C_2, \cdots, C_k$ are framed smooth simple closed curves. Fur-thermore, they are disjoint.

DEFINITION 5.22 (Link). The union $L = C_1 \cup C_2 \cup \cdots \cup C_k$ of mutually disjoint, smooth simple closed curves $C_1, C_2, \cdots, C_k$ in S^3 is called a *link*.

A link consisting of a single simple closed curve C_1 is called a *knot*. A link is a disjoint union of finitely many knots.

A Heegaard diagram without 1-handles of a closed 4-manifold M is a *framed link* in S^3.

For mutually disjoint simple closed curves C_1, C_2 in S^3, if they are oriented, an integer called the *linking number* between C_1 and C_2 is determined. The linking number is denoted by

$$\mathrm{Link}(C_1, C_2).$$

The value of $\mathrm{Link}(C_1, C_2)$ remains unchanged if C_1 moves con-tinuously in $S^3 - C_2$. The same can be said if the roles of C_1 and C_2 are switched.

Also,

$$\text{Link}(C_1, C_2) = \text{Link}(C_2, C_1).$$

We leave the definition of linking number to other textbooks (for example, [3], §4.5), and here we describe how to compute $\text{Link}(C_1, C_2)$ from a "picture" of C_1 and C_2.

Suppose that a picture of C_1 and C_2 is drawn as in Figure 5.14. In this picture we look at the double points of C_1 and C_2, where they overlap. We ignore the double points of either component by itself (self-crossings). Give the value $+1$ to a double point if it looks like Figure 5.15 left with respect to the orientations of C_1 and C_2, and the value -1 if it looks like Figure 5.15 right. If we add up all such values over all double points between C_1 and C_2, then we obtain an even integer, which is two times $\text{Link}(C_1, C_2)$.

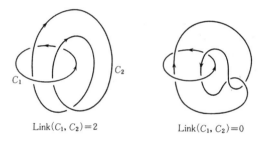

$$\text{Link}(C_1, C_2) = 2 \qquad \text{Link}(C_1, C_2) = 0$$

FIGURE 5.14. A sample calculation of $\text{Link}(C_1, C_2)$

FIGURE 5.15. A positive double point and a negative double point

The linking number depends on the choice of orientation of S^3, and the computational method we described here corresponds to the choice of orientation of "right-hand system." Figure 5.14 depicts examples of calculations of $\text{Link}(C_1, C_2)$ by this method.

LEMMA 5.23. *For a framed link $C_1 \cup C_2 \cup \cdots \cup C_k$ in S^3, an equivalence class of frames can be specified by assigning an integer a_i to each C_i.*

PROOF. Give an arbitrary orientation to C_i. For a given frame $\{\mathbf{v}_1, \mathbf{v}_2\}$ along C_i, let C_i' be a push-off of C_i in the direction of the frame. That is, C_i' is obtained from C_i by pushing it in the direction of $\mathbf{v}_1$. The curves C_i and C_i' run together "in parallel," and do not intersect each other. Give the orientation to C_i' which is parallel to that of C_i. Then the linking number

$$\text{Link}(C_i, C_i')$$

is determined, and assign this integer a_i to C_i. The value of a_i does not depend on the choice of orientation of C_i arbirarily picked at the beginning. (This is because the orientation of C_i' is also reversed if we reverse that of C_i.)

Conversely, if an integer a_i is given, it is clear that a vector field $\mathbf{v}_1$ can be constructed along C_i in such a way that $\text{Link}(C_i, C_i') = a_i$. Define $\mathbf{v}_2$ by rotating $\mathbf{v}_1$ counterclockwise by $90°$.

This completes the proof of Lemma 5.23.   □

EXAMPLE 5.24 (Heegaard diagram of $\mathbb{C}P^2$). We use the same notation as in Example 3.10 in Chapter 3. Points of $\mathbb{C}P^2$ are represented by $[z_1, z_2, z_3]$. For a non-zero complex number α, $[\alpha z_1, \alpha z_2, \alpha z_3]$ and $[z_1, z_2, z_3]$ represent the same point.

The critical point of index 0 of the Morse function $\mathbb{C}P^2 \to \mathbb{R}$ given in Example 3.10 is $[1, 0, 0]$. As an upper disk corresponding to this point, we take

$$\Delta^4 = \{ [1, z_2, z_3] \mid |z_2|^2 + |z_3|^2 \le 1 \}.$$

Furthermore, the critical point of index 2 is $[0, 1, 0]$, and the cocore of the corresponding 2-handle is given by

$$\Delta^2 = \{ [z_1, 1, 0] \mid |z_1| \le 1 \}.$$

The intersection $\Delta^4 \cap \Delta^2$ is

$$C = \{ [1, \exp(\sqrt{-1}\,\theta), 0] \mid \theta \in \mathbb{R} \}.$$

Therefore, the Heegaard diagram $(\Delta^4; C)$ of $\mathbb{C}P^2$ consists of the 4-disk Δ^4 and a standard circle C in the boundary $S^3 = \partial\Delta^4$ (an unknotted circle, known as the *trivial knot* or *unknot*).

The frame of C is given by the integer 1. This can be computed directly, or, one can prove it from Exercise 4.2 in Chapter 4 and the following Proposition 5.25. Thus, the framed link in Figure 5.16 has been obtained as a Heegaard diagram of $\mathbb{C}P^2$.

The manifold $-\mathbb{C}P^2$, obtained from $\mathbb{C}P^2$ by giving the opposite orientation to the one determined naturally from the complex structure, has a Heegaard diagram with the frame -1 along C.

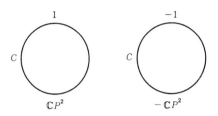

FIGURE 5.16. Heegaard diagrams of $\mathbb{C}P^2$ and $-\mathbb{C}P^2$

PROPOSITION 5.25. *The homology group $H_2(M)$ of a closed 4-manifold M, which has a handle decomposition without 1- or 3-handles, is a free abelian group. The intersection form*

$$I : H_2(M) \times H_2(M) \to \mathbb{Z}$$

can be computed from a Heegaard diagram $(D^4; C_1, C_2, \cdots, C_k)$ as follows. Give an arbitrary orientation to each closed curve C_i. An oriented C_i uniquely determines an element x_i of $H_2(M)$. Namely, the homology class represented by the core of the 2-handle, whose attaching sphere (circle) is C_i, is x_i. The set $\{x_1, x_2, \cdots, x_k\}$ forms a basis of $H_2(M)$. Furthermore, the following relation holds:

$$I(x_i, x_j) = \begin{cases} \text{Link } (C_i, C_j) & (if\ i \neq j), \\ a_i & (if\ i = j). \end{cases}$$

The fact that $H_2(M)$ is a free abelian group when a handle decomposition of M does not contain 1- or 3-handles can be seen from Theorem 4.18 in Chapter 4 and the definition of a chain complex. The cells of cores of 2-handles give generators. The relation between intersection forms and linking numbers is a well-known, fundamental fact, and a proof is omitted here.

EXAMPLE 5.26. Figure 5.17 represents a Heegaard diagram of $S^2 \times S^2$.

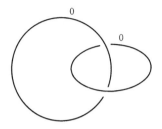

FIGURE 5.17. A Heegaard diagram of $S^2 \times S^2$

(c) Kirby calculus

Kirby [6] considered handlebodies of the form

$$N_L = D^4 \cup (h_1^2 \sqcup h_2^2 \sqcup \cdots \sqcup h_k^2)$$

obtained from D^4 by attaching 2-handles along L, starting from framed links $L = (C_1, a_1) \cup (C_2, a_2) \cup \cdots \cup (C_k, a_k)$ in S^3. This $(D^4; C_1, C_2, \cdots, C_k)$ is not necessarily a Heegaard diagram of a closed 4-manifold. Rather, Kirby's concern was the 3-manifolds $M_L^3 = \partial N_L$ that appear as the boundary.

THEOREM 5.27 (Lickorish [9]). *For any oriented, connected, closed 3-manifold M^3, there is a framed link $L = (C_1, a_1) \cup (C_2, a_2) \cup \cdots \cup (C_k, a_k)$ in S^3 such that M^3 is diffeomorphic to M_L^3 ($= \partial N_L$), with orientations taken into consideration. Here, as a convention, M_L^3 is endowed with the orientation which extends the right-hand orientation of S^3.*

According to Theorem 5.27, research on oriented closed 3-manifolds is reduced, at least in principle, to research on framed links in S^3.

Kirby considered the following moves I, II as the deformations to framed links L.

Move I: Replace L by L', where L' is the union $L \cup (C_0, \pm 1)$ of L and the trivial knot C_0 with a frame of $+1$ or -1, $(C_0, \pm 1)$. In this case, C_0 must be positioned far away from L and unlinked with L (cf. Figure 5.18).

Move II: Consider two connected components of L, say (C_1, a_1) and (C_2, a_2), and let C_1' be a push-off of C_1 in the direction of its frame a_1. Denote by $C_1' \#_b C_2$ the curve obtained from C_1' and C_2 by taking a connected sum along a "band" b as depicted in Figure 5.19. In this case, assume that the band b does not intersect L except at

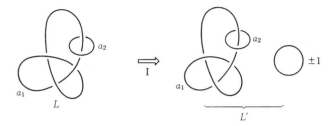

FIGURE 5.18. Move I

the ends, where it intersects C_1' and C_2 respectively. Replace C_2 by $C_1' \#_b C_2$. Finally, specify a frame of $C_1' \#_b C_2$ by the integer

$$(5.9) \qquad a_2' := a_1 + a_2 \pm 2 \operatorname{Link}(C_1, C_2),$$

where the choice of $+$ or $-$ depends on whether or not the orientation of $C_1' \#_b C_2$ matches the orientation parallel to C_1, when the orientation of $C_1' \#_b C_2$ is chosen in such a way that it matches the orientation of C_2 (cf. Figure 5.19).

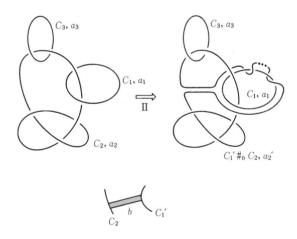

FIGURE 5.19. Move II

Move II corresponds to a handle slide of the 2-handle corresponding to C_2 over the 2-handle corresponding to C_1, in the handlebody N_L. Therefore, the diffeomorphism type of N_L does not change by performing a move II. In particular, the diffeomorphism type (with orientation considered) of the boundary 3-manifold M_L^3 remains unchanged as well.

A proposition similar to Proposition 5.25 holds for the intersection form on $H_2(N_L)$, and each component of the framed link corresponds to a 2-dimensional homology class of N_L. The difference from the case of closed 4-manifold is that I is not necessarily nondegenerate.

After a handle slide, the homology class corresponding to $C_1' \#_b C_2$ is $\pm x_1 + x_2$. The reason why we specified the integer a_2' determined by the formula (5.9) as the frame of $C_1' \#_b C_2$ is the following formula:

$$I(\pm x_1 + x_2, \pm x_1 + x_2) = I(x_1, x_1) + I(x_2, x_2) \pm 2I(x_1, x_2).$$

Move I alters N_L. However, it does not alter the diffeomorphism type (with orientation) of the boundary M_L^3.

THEOREM 5.28 (Kirby [6]). *For framed links L_1 and L_2 in S^3, a necessary and sufficient condition for $M_{L_1}^3$ and $M_{L_2}^3$ to be diffeomorphic 3-manifolds (with orientations considered) is that L_1 and L_2 are related by a finite number of operations that are either move I, move II, or their inverse operations.*

A framed link L in S^3 represents a 4-manifold as a Heegaard diagram, as described in the preceding section, and represents a 4-dimensional handlebody of the form N_L; but on the other hand, it is also considered to represent a 3-manifold of the form M_L^3 ($= \partial N_L$). When considered that way, L is called a *Kirby diagram* of the 3-manifold M_L^3. Framed links in Figure 5.16 (left) and Figure 5.17 represent $\mathbb{C}P^2$ and $S^2 \times S^2$, respectively, as Heegaard diagrams of closed 4-manifolds, but both represent S^3 as Kirby diagrams of 3-manifolds.

EXAMPLE 5.29 (Kirby diagrams of lens spaces). Let p, q be relatively prime integers such that $1 \le q < p$. Let

$$a_1 - \cfrac{1}{a_2 - \cfrac{1}{\ddots - \cfrac{1}{a_k}}} \qquad (a_i \ge 2)$$

be a continuous fraction expansion of p/q. Then a Kirby diagram of $L(p, q)$ is as shown in Figure 5.20. (Cf. [3], Figure 5.31.)

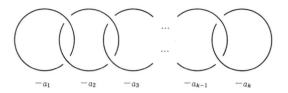

$-a_1$ $-a_2$ $-a_3$ $-a_{k-1}$ $-a_k$

FIGURE 5.20. A Kirby diagram of the lens space $L(p, q)$

Changing framed links by moves I, II, and their inverses is called *Kirby calculus* of framed links. The moves I, II, and their inverses are called *Kirby moves*.

Kirby calculus is often useful in proving that two 3-manifolds are actually diffeomorphic. The usefulness of Kirby's Theorem 5.28 was not clear when it was proved, since it does not give an algorithm for determining diffeomorphism types. Today, however, more than twenty years after the theorem was proved, its theoretical siginificance has become widely recognized.

Summary

5.1 The fundamental groups of cell complexes can be computed by van Kampen's theorem.

5.2 Closed surfaces can be classified completely by orientability and the Euler number.

5.3 Diffeomorphism types of oriented, connected, closed 3-manifolds can be described by Heegaard diagrams.

5.4 Diffeomorphism types of oriented, connected, closed 4-manifolds can be described by 4-dimensional Heegaard diagrams.

5.5 Diffeomorphism types of oriented, connected, closed 3-manifolds can be described also by framed links in S^3 (Kirby diagrams).

5.6 Kirby moves on Kirby diagrams do not alter diffeomorphism types of 3-manifolds represented by the diagrams.

Exercises

5.1 Suppose that three 1-handles, h_1^1, h_2^1, and h_3^1, are attached to the disk D^2. If all these 1-handles are attached in orientation-preserving ways, then the boundary $\partial(D^2 \cup h_1^1 \cup h_2^1 \cup h_3^1)$ is not connected. Prove this fact.

5.2 Prove that the handlebody depicted in Figure 5.7 is diffeomorphic to the one depicted in Figure 5.8.

5.3 Construct a cell decomposition of the 3-dimensional projective space $P^3 = e^0 \cup e^1 \cup e^2 \cup e^3$ such that the union of the first three cells $e^0 \cup e^1 \cup e^2$ is a cell decomposition of the project plane P^2.

A View from Current Mathematics

Morse thoery was established in the 1930's. Since then, Morse theory has enjoyed exciting developments where finite and infinite dimensions cross each other. In this series, however, Morse theory is discussed in two separate volumes, one for finite dimensions and the other for the infinite, for convenience. This is the volume which discusses finite dimensions.

The most important theory in finite-dimensional Morse theory is the theory of handlebodies, established by Smale. Consequently, the main topic of this book is handlebodies. During the 1960's, when higher-dimensional differential topology was developed, two significant theories were created based on handlebody theory: the h-cobordism theorem and surgery theory. It is not an overstatement to say that most of the results in higher-dimensional differential topology have been obtained by these two theories, together with the Kirby-Siebenmann theory, announced in 1969.

After the 1970's, low-dimensional manifolds have especially attracted attention. The theories for low-dimensional manifolds, which are essentially equivalent to handlebody theory, had been discussed at the beginning of the twentieth century by people such as Poincaré, Heegaard and Dehn. It was, however, after the 70's that work by these people started attracting more attention. In particular, the contribution of "Kirby calculus on framed links" to the explicit visualization of low-dimensional manifolds is significant. It combines results by Lickorish and Wallace in the 60's and work by Cerf.

Furthermore, Kirby calculus made the relationship stronger than ever before between low-dimensional manifold theory and knot theory. Recently, both theories seem to have been unified, in the sense that results in knot theory have been generalized to 3-manifold theory through Kirby calculus, and vice versa.

There are infinitely many ways to tie a knot in a circle in 3-dimensional space, and such complexity indicates that low-dimensional manifold theory is not straightforward. The handlebody theories for low-dimensional manifolds enable us to visualize the manifolds explicitly, but there are often difficulties in dealing with low-dimensional handles directly. For example, it is a tough road to develop 3-manifold theory using Heegaard decompositions.

More recently, instead of directly using Heegaard decompositions or Kirby diagrams, theories of 3-manifold invariants that are constructed using these diagrams have been developed extensively. The invariants discovered this way include Casson invariants, Kohno invariants, various quantum invariants, and Ohtsuki's finite type invariants.

Interestingly, the origin of these invariants lies, at least in principle, in the infinite-dimensional Morse theory on functionals (Chern-Simons functionals) on the space of all "connections."

Not only in dimension 3, but also in 4-manifold theory, very deep theories have been constructed, starting from applications of gauge theory by Donaldson (at the beginning of the 1980's). The shock of discovery of exotic 4-spaces (1983) is still fresh in one's memory; their existence was proved by combining Freedman's theory on 4-dimensional topological manifolds and Donaldson theory. Donaldson theory is also a Morse theory on functionals (Yang-Mills functionals) on infinite-dimensional spaces formed by all $SU(2)$-connections. Furthermore, more recently, Seiberg-Witten theory significantly simplified Donaldson theory.

It could be said that Morse theories of finite and infinite dimensions have started moving along together again, at the end of the twentieth century.

The original plan was to devote the last section of this book to the topology of algebraic surfaces. However, the number of pages significantly exceeded the planned amount, and the section on algebraic surfaces had to be cut.

In that section I was planning to discuss the theory of Lefschetz fibrations. This should be called a Morse theory in the world of complex numbers, and is important in the study of algebraic surfaces (which are a special type of 4-manifolds). In particular, monodromies around singular fibers play a role in combining 4-manifold theory, mapping class groups of 2-dimensional closed surfaces, and combinatorial group theory. There are many topics left to be investigated on these topics, and they will hopefully be discussed in the future elsewhere. (Recently Gompf and Stipsicz's book [3] has been published, and Chapter 8 of their book deals with this subject.)

Finally, I would like to add my personal recollections about the very last year of Professor Morse's life. He passed away in 1977, and I was staying at the Institute at Princeton at the time. Professor Morse was over eighty years old when I met him before he passed away. I remember him playing a piano at a party held at his house. I visited his office, talked to him about my research, and asked him for some reprints of his papers. One day in the late afternoon, when I was standing by the mailboxes in the Institute, I heard Mrs. Morse calling Professor Morse, "Marston! Marston!" I didn't see her, but her voice echoed in the hallway, thick in dusk. A little while after the incident, Professor Morse died. A funeral was held at the church on the campus of Princeton University. I saw Mrs. Morse greeting attendants of the funeral at the exit, and suddenly, I could not hold back my tears.

I would like to dedicate this book to the memory of Professor Morse.

Answers to Exercises

Chapter 1

1.1 Decompose the standard 2-dimensional sphere S^2 into the northern and southern hemispheres: $S^2 = D_- \cup D_+$. Let $H_2 : D_2 \to D_-$ be an arbitrary diffeomorphism; then Lemma 1.20 implies that the composite $(H_2|\partial D_2) \circ h : \partial D_1 \to \partial D_- = \partial D_+$ extends to a diffeomorphism $H_1 : D_1 \to D_+$. By gluing two difeomorphisms H_1 and H_2, then, we obtain a single diffeomorphism $H : D_1 \cup_h D_2 \to S^2$.

1.2 First, put the polar coordinates (r, θ) on D^2. The portion corresponding to $r = 1$ is the circle S^1 and the portion coresponding to $r = 0$ is the origin $\mathbf{0}$. Then we can simply define $H : D^2 \to D^2$ by

$$H(r, \theta) = (r, h(\theta)).$$

If h is a homeomorphism, then so is H. However, even if the original $h : S^1 \to S^1$ is smooth, in general $H : D^2 \to D^2$ may not be differentiable at the origin. Therefore, a diffeomorphism $h : S^1 \to S^1$ cannot necessarily be extended to a diffeomorphism $H : D^2 \to D^2$ by this method. (The argument we used here to extend a homeomorphism of the boundary of D^2 to a homeomorphism of D^2 can be applied to the general n-dimensional disk D^n, and is called Alexander's trick. Compare with Exercise 3.1.)

1.3 Impose the angular coordinate θ on S^1. By composing with the flipping diffeomorphism of D^2 if necessary, we may assume that h does not change the orientation of S^1. Then we can assume that $\dfrac{dh}{d\theta}(\theta) > 0$. First we extend $h : S^1 \to S^1$ to a diffeomorphism $H' : S^1 \times [0, 1] \to S^1 \times [0, 1]$ of an annulus. For that purpose, we can simply set

$$H'(\theta, t) = (th(\theta) + (1 - t)\theta, \ t), \quad \forall(\theta, t) \in S^1 \times [0, 1].$$

Since the determinant of the Jacobian of H' is alway positive, H' is a diffeomorphism. The original S^1 is identified with $S^1 \times \{1\}$.

Since H' restricts to $S^1 \times \{0\}$ as the identity mapping, H' can be extended to a diffeomorphism of $S^1 \times [0,1] \cup D^2$ by gluing D^2 along $S^1 \times \{0\}$ and extending H' as the identity on D^2, and thus h extends to $S^1 \times [0,1] \cup D^2$ as well. (I learned this simple argument from Koichi Yano.)

1.4 At any point on a torus, (θ, ϕ) can be used as a local coordinate system. By solving

$$\frac{\partial f}{\partial \theta} = -(R + r\cos\phi)\sin\theta = 0, \quad \frac{\partial f}{\partial \phi} = (-r\sin\phi)\cos\theta = 0,$$

we obtain four critical points $(\theta, \phi) = (0,0), (0,\pi), (\pi,0), (\pi,\pi)$. By computing the Hessian H_f at each critical point, we see that these critical points are non-degenerate, and that the indices of $(0,0)$, $(0,\pi)$, $(\pi,0)$, (π,π) are $2, 1, 1, 0$, respectively.

Chapter 2

2.1 Take different local coordinate systems $(x_1, \ldots, x_m)$, $(y_1, \ldots, y_m)$ about the point p_0. From the formula of changing coordinates for partial derivatives

$$\frac{\partial f}{\partial x_i}(p_0) = \sum_{j=1}^{m} \frac{\partial y_j}{\partial x_i} \frac{\partial f}{\partial y_j}(p_0),$$

we see that if p_0 is a critical point of f $\left(\dfrac{\partial f}{\partial y_1}(p_0) = \right.$ $\cdots = \dfrac{\partial f}{\partial y_m}(p_0) = 0\Big)$ with respect to $(y_1, \ldots, y_m)$, then p_0 is a critical point of f $\left(\dfrac{\partial f}{\partial x_1}(p_0) = \cdots = \dfrac{\partial f}{\partial x_m}(p_0) = 0\right)$ with respect to $(x_1, \ldots, x_m)$ as well. The same holds if the roles of the two coordinate systems are switched, so that p_0 being a critical point of f does not depend on the choice of local coordinate systems.

2.2 There are only two critical points, $(0, \ldots, 0, \pm 1)$. The index of $(0, \ldots, 0, -1)$ is 0, and that of $(0, \ldots, 0, 1)$ is $m-1$. See also Example 1.15 in Chapter 1.

2.3 Let (p, q) be an arbitrary point of the product $M \times N$, and let $(x_1, \ldots, x_m)$ and $(y_1, \ldots, y_n)$ be local coordinate systems about p and q, respectively. Then we can take $(x_1, \ldots, x_m, y_1, \ldots, y_n)$ as a local coordinate system about (p,q) in $M \times N$. Since

$$\frac{\partial F}{\partial x_i} = \left(\frac{\partial f}{\partial x_i}\right)(B+g) \quad \text{and} \quad \frac{\partial F}{\partial y_j} = (A+f)\left(\frac{\partial g}{\partial y_j}\right),$$

by taking A and B large enough in M and N respectively in such a way that $A > |f|$ and $B > |g|$, the following two conditions become equivalent:

$$\frac{\partial F}{\partial x_i}(p,q) = \frac{\partial F}{\partial y_j}(p,q) = 0 \quad (i=1,\ldots,m,\ j=1,\ldots,n)$$

and

$$\frac{\partial f}{\partial x_i}(p) = 0 \ (i=1,\ldots,m) \quad \text{and} \quad \frac{\partial g}{\partial y_j}(q) = 0 \ (j=1,\ldots,n).$$

Therefore, the only critical points of F are of the form (p_0,q_0), where p_0 and q_0 are critical points of f and g respectively. The index of (p_0,q_0) is the sum of the indices of p_0 and q_0. (Compute this by representing f and g in standard forms in the neighborhoods of p_0 and q_0.)

2.4 The conditions $\dfrac{\partial f}{\partial \theta_i} = 0$ and $\theta_i = 0, \pi$ are equivalent (ignoring any difference by integral multiples of 2π). Therefore, the critical points are $(\varepsilon_1, \varepsilon_2, \ldots, \varepsilon_m)$ (where $\varepsilon_i = 0, \pi$). We omit a proof that these critical points are non-degenerate. The index of $(\varepsilon_1,\ldots,\varepsilon_m)$ is equal to the number of ε_i's with $\varepsilon_i = 0$.

Chapter 3

3.1 Let D^m be the m-dimensional disk of radius 1, with center $\mathbf{0}$. Let (r,x) be the point of D^m at a distance r from $\mathbf{0}$, lying on the straight line segment connecting $\mathbf{0}$ and a point x of S^{m-1}. Define $\bar{h}: D^m \to D^m$ by $\bar{h}(r,x) = (r, h(x))$.

3.2 Although this problem is intuitively obvious, it is not that easy when one tries to write it down in explicit expressions. There are various methods, but here we construct an isotopy $\{h_t\}_{t\in J}$ in the following steps. First, pick a smooth function $f(x)$ of a single variable satisfying

$$f(x) = \begin{cases} 1 & \left(|x| < \frac{1}{3}\right), \\ 0 & \left(|x| > \frac{1}{2}\right), \end{cases} \quad 0 \le f(x) \le 1.$$

(See Section 1 of [17] for the existence of such an $f(x)$.)

For a sufficiently small $\varepsilon > 0$, the function

$$f_\varepsilon(x) = \varepsilon f(x) + x$$

is strictly increasing with respect to x, and $f_\varepsilon(x) = x$ if $|x| > 1/2$. Furthermore, $f_\varepsilon(0) = \varepsilon$. Choose yet another smooth non-decreasing function with a single variable $\rho_\varepsilon(x)$ such that

$$\rho_\varepsilon(x) = \begin{cases} 0 & \left(x < \frac{\varepsilon}{2}\right), \\ 1 & (x > \varepsilon). \end{cases}$$

(See Section 1 of [17] again for the existence of such a $\rho_\varepsilon(x)$.)

With the above preliminaries we define a smooth function $g_\varepsilon(x_1, \cdots, x_{k-1}, x_k)$ with k variables as follows:

$$
\begin{aligned}
&g_\varepsilon(x_1, \cdots, x_{k-1}, x_k) \\
&= (1 - \rho_\varepsilon(x_1^2 + \cdots + x_{k-1}^2))f_\varepsilon(x_k) + \rho_\varepsilon(x_1^2 + \cdots + x_{k-1}^2)x_k.
\end{aligned}
$$

With this definition for g_ε, we see that

$$g_\varepsilon(x_1, \cdots, x_{k-1}, x_k) = \begin{cases} f_\varepsilon(x_k) & (\text{if } x_1^2 + \cdots + x_{k-1}^2 < \frac{\varepsilon}{2}), \\ x_k & (\text{if } x_1^2 + \cdots + x_{k-1}^2 > \varepsilon). \end{cases}$$

from the property of ρ_ε. Furthermore, we see that

$$\text{if } |x_k| > \frac{1}{2}, \text{ then } g_\varepsilon(x_1, \cdots, x_{k-1}, x_k) = x_k,$$

from the property of $f_\varepsilon(x_k)$. The function $g_\varepsilon(x_1, \cdots, x_{k-1}, x_k)$ is strictly increasing with respect to x_k, and $g_\varepsilon(0, \cdots, 0, 0) = f_\varepsilon(0) = \varepsilon$. Define a diffeomorphism $f : D^k \to D^k$ by

$$
\begin{aligned}
&h(x_1, \cdots, x_{k-1}, x_k) \\
&= (x_1, \cdots, x_{k-1}, g_\varepsilon(x_1, \cdots, x_{k-1}, x_k)).
\end{aligned}
$$

Then, from the property of $g_\varepsilon(x_1, \cdots, x_{k-1}, x_k)$ mentioned above, h is the identity map id on a neighborhood of ∂D^k (with a sufficiently small ε), and furthermore,

$$h(0, \cdots, 0, 0) = (0, \cdots, 0, \varepsilon).$$

Using again the non-decreasing function ρ_ε we already picked, define an isotopy $\{h_t\}_{t \in J}$ by

$$
\begin{aligned}
&h_t(x_1, \cdots, x_{k-1}, x_k) \\
&= (x_1, \cdots, x_{k-1}, \rho_\varepsilon(t)g_\varepsilon(x_1, \cdots, x_{k-1}, x_k) + (1 - \rho_\varepsilon(t))x_k).
\end{aligned}
$$

Then, certainly,
$$h_t = \begin{cases} \text{id} & (\text{if } t \leq 0), \\ h & (\text{if } t \geq \varepsilon). \end{cases}$$

For $t = 1$, h_1 coincides with h, which sends the origin $(0, \cdots, 0)$ of D^k to $(0, \cdots, 0, \varepsilon)$.

So far $\varepsilon > 0$ has been a sufficiently small number, and next we construct an isotopy such that
$$h_1(0, \cdots, 0) = (0, \cdots, 0, a)$$
for any positive number a with $\varepsilon < a < 1$. For this purpose let $\delta > 0$ be a sufficiently small positive number, and choose a non-decreasing function $\sigma(x)$ in such a way that
$$\sigma(x) = \begin{cases} \varepsilon/a & (x < a + \delta), \\ 1 & (x > a + 2\delta). \end{cases}$$

Then define a diffeomorphism $H : D^k \to D^k$ by
$$H(x_1, \cdots, x_k) = (\sigma(||x||)x_1, \cdots, \sigma(||x||)x_k),$$
where $||x||^2 = x_1^2 + \cdots + x_k^2$. The map H is the identity in a neighborhood of ∂D^k, and
$$H(0, \cdots, 0, a) = (0, \cdots, 0, \varepsilon).$$

Define the isotopy $\{H^{-1} \circ h_t \circ H\}_{t \in J}$ by composing the originally constructed isotopy $\{h_t\}_{t \in J}$ and H; then it is the identity map for $t \leq 0$, and $H^{-1} \circ h_1 \circ H$ sends the origin $(0, \cdots, 0)$ of D^k to $(0, \cdots, 0, a)$ for $t \geq 1$.

We obtain an isotopy which sends the origin to any point p_1 in the interior of D^k by making a composition of the above constructed isotopy and rotations of D^k. Furthermore, we obtain an isotopy which sends p_1 to p_2 for arbitrary points p_1 and p_2 in the interior of D^k, by making a composition of such an isotopy and the inverse of the other.

3.3 Let us solve this problem using the following trick for approximating continuous functions by smooth ones (see §4 of [17]). First, consider a smooth function σ of a single variable defined on $\mathbb{R}$ with the following four properties (see Figure A.1):

(i) $\sigma(t) \geq 0$.

(ii) If $|t| > \varepsilon$, then $\sigma(t) = 0$, where $\varepsilon > 0$ is sufficiently small.

(iii) $\displaystyle\int_{-\infty}^{\infty} \sigma(t)dt = 1$.

(iv) $\sigma(t) = \sigma(-t)$.

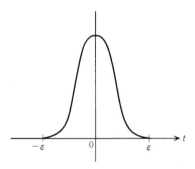

FIGURE A.1. The graph of $\sigma(t)$

Such a function is obtained, for example, from the function $h(t)$ in Section 1 of [17] by setting $c_\varepsilon(t) = h(t/\varepsilon)$ and then multiplying it by $\{1/\int_{-\infty}^{\infty} c_\varepsilon(t)dt\}$. For a continuous function $f(t)$ on $\mathbb{R}$, the function $g(x)$ defined by

$$g(x) = \int_{-\infty}^{\infty} f(t)\sigma(x - t)dt$$

is an avarage of $f(t)$ with weight on the domain $|t - x| < \varepsilon$ of t. Therefore, the smaller ε is, the better approximation $g(x)$ is for $f(x)$. Furthermore, if $f(t)$ is strictly increasing, then so is $g(x)$. Note that, from the above definition, $g(x)$ is smooth with respect to x even if f is not differentiable. Moreover, from the symmetry (the condition (iv)) of σ, we see the following. Namely, for a real number q, if $f(t)$ is a linear function ($f(t) = Ax + B$) on the domain $|t - q| < 2\varepsilon$ around q, then $g(x)$ is also a linear function of the same form ($g(x) = Ax + B$) on the domain $|t - q| < \varepsilon$.

Let us go back and solve Exercise 3.3 as an application of the above general method. The notations in the following are the same as those in Exercise 3.3. Let $f_0(t)$ be a "piecewise linear" function obtained by gluing linear functions, with the following three properties:

(i) $f_0(t)$ is strictly increasing with respect to t.

(ii) $f_0(t) = t$ if $t < c + 2\varepsilon$ or $t > d - 2\varepsilon$, where ε is the positive number used in defining σ above.

(iii) $f_0(t) = t + a - q_1$ if $|t - q_1| < 2\varepsilon$.

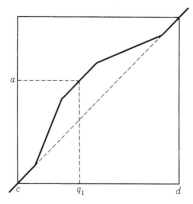

FIGURE A.2. The graph of $f_0(t)$

The graph of $f_0(t)$ looks like Figure A.2. Define a smooth function $g_0(x)$ which approximates this f_0 by

$$g_0(x) = \int_{-\infty}^{\infty} f_0(t)\sigma(x-t)dt.$$

Then $g_0(x)$ is certainly smooth, and has the following three properties:

(i) $g_0(x)$ is strictly increasing, and $g_0(x)$ increases from c to d as x increases from c to d.

(ii) $g_0(x) = x$ if $x < c + \varepsilon$ or $x > d - \varepsilon$.

(iii) $g_0(x) = x + a - q_1$ if $|x - q_1| < \varepsilon$. In particular, $g_0(q_1) = a$.

Next, construct a function $g_1(x)$ with similar properties by replacing q_1 by q_2 and a by b. In particular, $g_1(x)$ satisfies $g_1(q_2) = b$.

Choose a smooth non-decreasing function $\rho(s)$ such that

$$\rho(s) = \begin{cases} 0, & s \leq 0, \\ 1, & s \geq 1. \end{cases}$$

A desired function $G(x, s)$ can now be defined by

$$G(x, s) = (1 - \rho(s))g_0(x) + \rho(s)g_1(x).$$

3.4 As we computed in Example 3.11, the Hessian at the critical point in question is a diagonal matrix of size $\dfrac{m(m-1)}{2} \times \dfrac{m(m-1)}{2}$. The "$(i, k)$-th" diagonal entry is

$$-c_i\varepsilon_i - c_k\varepsilon_k.$$

Assume that $i < k$. From the facts that $1 < c_1 < c_2 < \cdots < c_m$ and $\varepsilon_i = \pm 1$, it follows that this diagonal entry is negative if $\varepsilon_k = 1$ (no matter what the sign of ε_i is). For such a k, there are $k - 1$ of the pairs (i, k). Therefore, if we arrange such k's with $\varepsilon_k = 1$ from small ones and have

$$k_1, \; k_2, \; \cdots, \; k_n,$$

then the index of the Hessian is given by

$$(k_1 - 1) + (k_2 - 1) + \cdots (k_n - 1).$$

Chapter 4

4.1 $H_0(P^2) \cong \mathbb{Z}$ and $H_1(P^2) \cong \mathbb{Z}_2$. For $i \geq 2$, $H_i(P^2) \cong \{0\}$.

4.2 This can be proved by non-degeneracy of the intersection form.

4.3 The Betti numbers $b_0(M), b_1(M), \cdots, b_m(M)$ of M, where $m = \dim M$, satisfy $b_i(M) = b_{m-i}(M)$ $(i = 0, 1, \cdots, m)$, due to Poincaré duality. This fact, together with the Euler-Poincaré formula $\chi(M) = \sum_{i=1}^{m} (-1)^i \, b_i(M)$, implies that $\chi(M) = 0$ if m is odd. (It is known that $\chi(M) = 0$ if M is a closed manifold of odd dimension, even if M is not orientable.)

Chapter 5

5.1 If the feet of h_1 do not cross those of h_2 or h_3 at all, then the boundary $\partial(D^2 \cup h_1 \cup h_2 \cup h_3)$ is not connected. Therefore, we may assume that a foot of h_1 crosses a foot of at least one 1-handle (say, h_2). Then the boundary $\partial(D^2 \cup h_1 \cup h_2)$ is connected (and is a circle). Hence the feet of h_3 can be made uncrossed with the feet of h_1 or h_2 by sliding the feet of the third 1-handle h_3. Then $\partial(D^2 \cup h_1 \cup h_2 \cup h_3)$ is not connected.

5.2 See Figure A.3.

5.3 Regard the 3-sphere S^3 as the unit sphere $\{(x, y, z, w) \mid x^2 + y^2 + z^2 + w^2 = 1\}$ in the 4-dimensional space. If we decompose S^3 into two 0-cells $e_\pm^0 = (\pm 1, 0, 0, 0)$, two 1-cells $e_\pm^1 = \{(x, y, 0, 0) \mid x^2 + y^2 = 1, \pm y > 0\}$, two 2-cells $e_\pm^2 = \{(x, y, z, 0) \mid x^2 + y^2 + z^2 = 1, \pm z > 0\}$, and two 3-cells $e_\pm^3 = \{(x, y, z, w) \mid x^2 + y^2 + z^2 + w^2 = 1, \pm w > 0\}$, then this decomposition gives a desired decomposition of P^3 on the quotient.

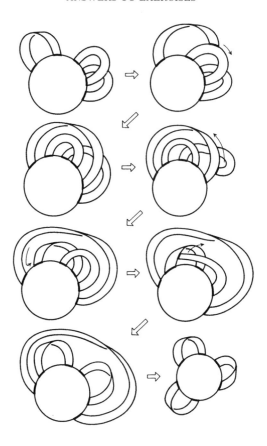

FIGURE A.3

Bibliography

[1] J. Cerf, Sur les difféomorphismes de la sphère de dimension trois ($\Gamma_4 = 0$), Lecture Notes in Math. **53**, Springer-Verlag, 1968.

[2] M. H. Freedman, The topology of four-dimensional manifolds, *J.Diff. Geom.*, **17** (1982), 337-453.

[3] R. E. Gompf and A. I. Stipsicz, *4-manifolds and Kirby calculus*, Graduate Studies in Mathematics **20**, Amer. Math. Soc., 1999.

[4] V. Guillemin and A. Pollack, *Differential topology*, Prentice-Hall, 1974.

[5] M. A. Kervaire and J. W. Milnor, Groups of homotopy spheres I, *Ann. of Math.*, **77** (1963), 504-537.

[6] R. C. Kirby, A calculus for framed links in S^3, *Invent. Math.*, **45** (1978), 35-56.

[7] S. Lang, *Differential and Riemannian manifolds*, GTM **160**, Springer-Verlag, 1995.

[8] F. Laudenbach and V. Poenaru, A note on 4-dimensional handlebodies, *Bull. Soc. Math. France,* **100** (1972), 337-347.

[9] W. B. R. Lickorish, A representation of orientable combinatorial 3-manifolds, *Ann. of Math.*, **76** (1962), 531-540.

[10] W. S. Massey, *A basic course in algebraic topology*, GTM **127**, Springer-Verlag, 1991.

[11] Y. Matsushima, *Differentiable manifolds*, Shouka-bou, Tokyo, 1965 (in Japanese). English translation: *Differentiable manifolds*, Pure and Applied Mathematics, **9**. Marcel Dekker, 1972.

[12] B. Mazur, Morse theory, *Differential and combinatorial topology, A symposium in honor of Marston Morse* (ed. by S. S. Cairns), Princeton Univ. Press, 1965, 145-165.

[13] J. Milnor, On manifolds homeomorphic to the 7-sphere, *Ann. of Math.*, **64** (1956), 399-405.

[14] J. Milnor, *Lectures on the h-cobordism theorem*, Princeton Univ. Press, 1965.

[15] J. Milnor, *Topology from the differentiable viewpoint*, U.P. of Virginia, 1965.

[16] J. M. Montesinos, Heegaard diagrams for closed 4-manifolds, *Geometric topology* (ed. by J. C. Cantrell) Academic Press, 1979, 219-237.

[17] J. R. Munkres, *Elementary differential topology*, Annals of Mathematics Studies **54**, Princeton Univ. Press, 1963.

[18] R. Palais, Extending diffeomorphisms, *Proc. Amer. Math. Soc.*, **11** (1960), 274-277.

[19] W. Rudin, *Principles of mathematical analysis*, 3rd ed., McGraw-Hill, 1976.

[20] I. Satake, *Linear algebra*, Shouka-bou, Tokyo, 1958 (in Japanese). English translation: *Linear algebra*, Pure and Applied Mathematics, **29**. Marcel Dekker, 1975.

[21] H. Sato, *Algebraic topology*, "Iwanami Kohza, Gendai suugaku no kiso"(Iwanami series in modern mathematics), 1996 (in Japanese). English translation: *Algebraic topology: An intuitive approach*, Translations of Mathematical Monographs **183**, Amer. Math. Soc., 1999.

[22] S. Smale, Diffeomorphisms of the 2-sphere, *Proc. Amer. Math. Soc.*, **10** (1959), 621-626.

[23] S. Smale, Generalized Poincaré's conjecture in dimensions greater than four, *Ann. of Math.*, **74** (1961), 391-406.

Recommended Reading

1. J. Milnor, *Morse theory,* Ann. Math. Studies, **51** (1963), Princeton U.P.

   The book starts from Morse theory of finite dimensions, and reaches the Morse theory for energy functionals on loop spaces of Lie groups. And at the end, a beautiful theorem called the Bott periodicity is proved, which is about the stable homotopy groups of Lie groups. The exposition is simple and easy to understand. This is a classical book with a well-deserved great reputation on Morse theory.

2. J. Milnor, *Lectures on the h-cobordism theorem,* Princeton U.P., 1965.

   The book explains the *h*-cobordism theorem, which is important in differential topology for higher-dimensional manifolds. The handlebody theory for manifolds is given in detail. This book is also easy to read and understand. Much of the treatment of handlebodies in the current book is based on this book of Milnor's.

3. R. C. Kirby, *The topology of 4-manifolds,* Lect. Notes in Math., **1374** (1989), Springer-Verlag.

   4-manifold theory is developed based on the common grounds between framed links and low-dimensional manifold theory. The exposition is sometimes intuitive, so first readers may have a hard time getting into it, although plenty of pictures are provided.

4. R. E. Gompf and A. I. Stipsicz, *4-manifolds and Kirby calculus,* Graduate Studies in Mathematics **20** (1999), Amer. Math. Soc.

   This book contains materials on 4-dimensional topology, from fundamental techniques to most recent results. Kirby calculus is emphasized significantly. Many exercises are also provided. This is quite a thick book, so it may be tough to read it through, but it could be used as a dictionary as well.

5. D. Rolfsen, *Knots and links,* Math. Lecture Series, **7** (1976), Publish or Perish, Inc.

   An introduction to knot theory, with a lot of pictures.

6. A. Kawauchi, *A survey of knot theory.* Translated and revised from the 1990 Japanese original by the author. Birkhäuser Verlag, Basel, 1996.

   This gives an overview on modern aspects of knot theory.

7. K. Fukaya, *Gauge theory and topology,* Springer-Verlag Tokyo, 1995 (in Japanese).

   This is mainly on the Morse theory of infinite dimensions, but a beautiful exposition on Morse theory is found in the introduction.

Index

Selected Titles in This Series

For a complete list of titles in this series, visit the
AMS Bookstore at **www.ams.org/bookstore/**.